THE BOULDER CANYON PROJECT

THE BOULDER CANYON PROJECT

Historical and Economic Aspects

By

PAUL L. KLEINSORGE
Assistant Professor of Economics and Business Administration
Oregon State College

With a Foreword by

ELIOT JONES
Professor of Public Utilities
Stanford University

Stanford University Press — Stanford University, California
London: Humphrey Milford — Oxford University Press

STANFORD UNIVERSITY PRESS
STANFORD UNIVERSITY, CALIFORNIA

LONDON: HUMPHREY MILFORD
OXFORD UNIVERSITY PRESS

THE BAKER AND TAYLOR COMPANY
55 FIFTH AVENUE, NEW YORK

THE MARUZEN COMPANY
TOKYO, OSAKA, KYOTO, SENDAI

PRINTED AND BOUND IN THE UNITED STATES
OF AMERICA BY STANFORD UNIVERSITY PRESS

To
My Mother
and to
The Memory of my Father

FOREWORD

A noteworthy feature of the past decade is the marked progress made in the development of the water powers of the country through vast federal projects. In the Southeast there is the Tennessee Valley Authority, which is promoting the utilization of the Tennessee River; in the Southwest the Boulder Canyon Project (Colorado River); in central California the Central Valley Authority (Sacramento and San Joaquin rivers); and in the Northwest the Grand Coulee and Bonneville projects (Columbia River).

These projects are known as multi-purpose projects. Not all of them have the same purposes, and the importance of the several purposes varies in the different projects; but the principal purposes are protection against floods, improvement of navigation, generation of electric energy, and provision of water for domestic use or irrigation. Other purposes include improvement of the land through fertilization and through prevention of soil erosion and salt-water intrusion, national defense, afforestation, conservation, sanitation, recreation, and preservation of wild life.

The first of the aforementioned projects in point of time, and among the first in size and importance, is the Boulder Canyon Project, which puts to effective use the waters of the mighty Colorado River. The watershed of this river includes seven states (Wyoming, Utah, Colorado, New Mexico, Arizona, Nevada, and California), and comprises about one-thirteenth of the area of the entire country. The Colorado has long been one of the most dangerous and destructive rivers in the world, as its flow varies greatly from season to season and from year to year, and as it is, with one possible exception, the largest silt-bearing stream in the world. The Boulder Canyon Project, and particularly Hoover (Boulder) Dam, will tame this wild river, and put it to work to generate electric energy and to supply water for domestic use and for irrigation.

To many persons the Boulder Canyon Project is synonymous

with Hoover (Boulder) Dam and its associated power plant and transmission lines; but the Project embraces much more than the dam, though that is a great engineering feat (and also a thing of beauty). Of great importance is the water-supply system of the Metropolitan Water District of Southern California, comprising the diversion dam near Parker, Arizona (one hundred and fifty-five miles south of Hoover Dam), the aqueduct across southern California together with the necessary pumping system, and the water distribution facilities. The Metropolitan Water District contracted to take 36 per cent of the electrical energy generated at Hoover Dam, the energy to be used principally in pumping Colorado River water across the mountains to reservoirs near Los Angeles. By contracting to take this large amount of energy the Metropolitan Water District made the Boulder Canyon Project feasible. Also of great importance is the irrigating system of the Imperial Irrigation District, comprising the Imperial Dam, a diversion dam one hundred and forty miles below Parker Dam, and the All-American Canal from Imperial Dam to the Imperial and Coachella valleys in southern California. The entire project, now nearing completion, will be at the date of its completion one of the largest flood-control, reclamation, water-storage, and power projects in the history of the world.

It is appropriate that this book, which is a scholarly and painstaking work, should be written by a Stanford man and published by the Stanford University Press, for the Stanford community naturally takes a particular pride in the Boulder Canyon Project and its principal dam. The pride is natural, since Herbert Hoover, a Stanford graduate and trustee, did more than any other man to bring the project to fruition; and since Dr. Ray Lyman Wilbur, President of the University, while Secretary of Interior had charge of the administration of the project, including the important duty of negotiating the power and water contracts. It was the successful negotiation of these contracts that gave the project unquestioned economic justification.

ELIOT JONES

STANFORD UNIVERSITY, CALIFORNIA
August 15, 1940

PREFACE

This book deals with the Boulder Canyon Project, the most important step yet taken toward the development of the Colorado River and the economic utilization of the great natural resources of that river's lower basin. It is not the first project undertaken in this area, nor is it likely to be the last; but it has led to the establishment of policies so vital to this section of the Southwest and of such interest to the nation as a whole that its outstanding importance is beyond question.

Special emphasis is placed here upon the historical and economic aspects in order to appraise the project from an economic point of view. The historical discussion includes much more than a chronological account of the progress of the construction activities. It presents a brief résumé of the earlier developments in the lower Colorado River Basin and of the problems encountered; it reviews the history of the Colorado River Compact, the Boulder Canyon legislation, and the important power and water contracts; it analyzes the Colorado River Compact and the Boulder Canyon Project Act and the arguments used to oppose them; it tells the story of the Colorado River Aqueduct, a related project; and it gives a detailed discussion of the legal controversies and of the court decisions. Attention is given also to the geological and engineering aspects, but with no attempt to provide a technical analysis.

The discussion of the economic aspects involves a study of the services of the project and their value in relation to the costs incurred. The most important services rendered are flood control and river regulation, water storage for agricultural use, water storage for domestic use (particularly in southern California), and the generation of electric power. In each case the gains to be realized are evaluated and possible alternative means of securing similar results are considered. Other services, such as the employment of large numbers of men during a period of depression, the creation of a new recreational area in the Southwest, and the provision of a foundation for future projects on the Colorado River, are also discussed. Such a study permits a fuller understanding of the economic justification of the project—and, for that matter, of any of the multi-purpose projects now under construction or in contemplation.

I am deeply indebted to numerous persons who encouraged and helped me in the preparation of this work. I am especially grateful to Professor Eliot Jones of Stanford University, whose careful guidance and helpful criticisms are largely responsible for whatever merit this study may possess. To President Donald M. Erb of the University of Oregon, and to Professor Edward Gene Nelson of Stanford University, I wish to express my appreciation for many valuable suggestions. And to Mr. John C. Page, Commissioner of the Bureau of Reclamation, Washington, D.C., I owe sincere thanks for information not otherwise available to me. I wish also to express my appreciation to Miss Minna Stillman and Miss Gladys Andrews for aid given in library research, and to Miss Elizabeth W. Breid for expert stenographic assistance.

PAUL L. KLEINSORGE

OREGON STATE COLLEGE
August 15, 1940

TABLE OF CONTENTS

LIST OF TABLES

LIST OF ILLUSTRATIONS

THE BOULDER CANYON PROJECT

Chapter I

CONDITIONS LEADING TO THE PROJECT

INTRODUCTION

The Boulder Canyon Project is nearing completion. The work on the great dam which holds back the waters of the Colorado River at Black Canyon is finished. The mammoth power plant at the dam, although not operating at capacity as yet, is supplying electric energy to the power markets of southern California. The All-American Canal will soon be used to carry a part of the waters of the Colorado River to the Imperial Valley in California; and within the year the Colorado River Aqueduct, a related project, will supply a number of the cities of southern California with water. With the completion of the project, one of the greatest flood-control, reclamation, water-storage, and power projects in the history of the world will be realized. The mighty Colorado River, which for centuries has been one of the world's most dangerous and destructive rivers, will have become a comparatively tame and docile stream held in check to generate electric energy, to reclaim the irrigable lands of the desert, and to serve the growing populations of southern California and of the Southwest.

Behind this great accomplishment lies a long train of physical, political, and economic developments. In spite of the general popularity of the development, the project is not without opponents who have argued long and vigorously against its construction. No one will deny that the development is a great feat of engineering, and that it is a monument to our modern civilization. Yet the arguments persist that the project is impractical, unwarranted, and even detrimental to social welfare under present economic conditions, or that another type of plan should have been followed to secure the best results. Actually a reasonable evaluation of the project's services cannot be made without a comprehensive background upon which to base conclusions. Therefore one of the purposes of this study is to provide the facts which make up such a background. The study will trace the steps leading to the project, will outline the problems involved, will analyze the arguments of the proponents and the opponents, and will attempt to show whether or not the benefits gained are a sufficient addition to the general welfare of the country to justify the project from an economic point of view.

DESCRIPTION OF THE COLORADO RIVER

COMPARATIVE SIZE

The Boulder Canyon Project was undertaken to develop, control, and use the services of the Colorado River.[1] This river with its tributaries drains an area of about 244,000 square miles, comprising approximately one-thirteenth of the area of the United States. The average flow of the river is about 17,000,000 acre-feet per annum, and its total length from the mouth in the Gulf of California to the headwaters in Colorado is nearly 1,500 miles, although a total length of 1,700 miles may be computed if the Green River, the longest tributary, is considered as its upper continuation.[2] The Colorado is surpassed in volume by a number of other rivers in the United States, but from the point of view of size alone its drainage area and length are sufficiently great to give it an important place among the river systems of this country.[3]

The Mississippi is the only river system in the United States which completely overshadows the Colorado in all respects.[4] It is the nation's greatest river, and with its tributaries it drains

[1] See Figure 1. See also map compiled by the Department of the Interior, United States Geological Survey, 1914.

[2] *Eighteenth Annual Report of the Reclamation Service (1918–1919)*, p. 391.

[3] See Table I.

[4] The St. Lawrence cannot be considered a United States river, because where it is not the border between the United States and Canada, it is entirely a Canadian stream. It should also be noted that through a great part of its length the Rio Grande is the boundary between Texas and Mexico.

approximately 41 per cent of the area of the country. The main stream of the Mississippi is about 2,560 miles long, but the distance from the mouth to the source of the Missouri River, the longest tributary, is about 4,200 miles, which exceeds the length of any other river in the world.[5]

TABLE I.—SOME IMPORTANT RIVERS OF THE UNITED STATES

	Average Yearly Volume (*Acre-Feet*)	Drainage Area (*Square Miles*)	Length (*Miles*)
St. Lawrence[a]	....	500,000	750
Mississippi	503,037,525[b]	1,240,000[c]	4,200
Ohio[d]	217,100,000[b]	201,700	....
Columbia[e]	151,000,000[f]	259,000	1,400
Missouri[d]	72,400,000[b]	527,100	....
Sacramento[g]	25,199,500	28,000	400
Colorado	17,000,000	244,000	1,700
Rio Grande[h]	4,846,000	248,000	2,000
Hudson[i]	....	13,366	300
Delaware[i]	....	10,100	410

[a] Angelo Heilprin and Louis Heilprin, editors, *Gazetteer of the World* (1922), p. 1610. Its volume of flow is estimated to be second in the world only to that of the Amazon.

[b] Water Supply Paper No. 345, *Contributions to the Hydrology of the United States* (1914), p. 77. The mean discharge of the Mississippi at its mouth is given as 695,000 second-feet, of the Missouri as 100,000 second-feet, and of the Ohio as 300,000 second-feet. See John Clayton Hoyt and Nathan Clifford Grover, *River Discharge* (1916), Table X, p. 202, for method of converting discharge in second-feet into runoff in acre-feet.

[c] House Document No. 798, 71st Congress, 3d Session, *Control of Flood in the Alluvial Valley of the Lower Mississippi River* (1931), Vol. I, p. 61.

[d] Tributary of the Mississippi.

[e] Federal Power Commission, *Report of the Federal Power Commission on the Uses of the Upper Columbia River* (1925), p. 6.

[f] Volume of flow at The Dalles, Oregon, over 100 miles upstream from Portland and 200 miles above the mouth of the river. This figure does not include, therefore, the flow of the tributaries below The Dalles, and is smaller than the flow of the river at its mouth.

[g] State of California Department of Public Works, Division of Engineering and Irrigation, Bulletin No. 5, *Flow in California Streams* (1923), p. 59.

[h] House Document No. 359, 71st Congress, 2d Session, *Report of the American Section of the International Water Commission, United States and Mexico* (1930), p. 25.

[i] Water Supply Paper No. 166, *Report of Progess of Stream Measurements for 1905*, pp. 16, 77.

These figures, when compared with corresponding statistics for the Colorado River, give the impression that the Colorado is a relatively unimportant stream; but such a comparison without qualification is misleading. The Hudson and Delaware rivers, extremely important rivers of the eastern coast from the

[5] Angelo Heilprin and Louis Heilprin, editors, *Gazetteer of the World* (1922), p. 1189.

economic point of view, are very small streams compared with the Colorado. The volume of flow of the St. Lawrence River is second in the world only to that of the Amazon; its drainage area is more than twice the size of the Colorado's; but unless the Great Lakes are included as a part of the river, its length from the heading in Lake Ontario to the mouth in the Gulf of St. Lawrence is only 750 miles.[6] The Rio Grande River likewise exceeds the Colorado both in length and drainage area but is far smaller in volume of flow. On the other hand, the Sacramento, largest river in California, is greater than the Colorado in volume but much smaller in length and drainage area. Finally, the Columbia River is the most important stream of western United States flowing into the Pacific Ocean, but it is not as long as the Colorado. The outstanding point of superiority of the Columbia is its enormous volume of flow.

These comparisons indicate that although the Colorado River is by no means the largest river in the United States, it is still one of the great rivers of the country. However, the importance of a stream is not derived from its size alone. An outstanding feature of the Colorado is the unusual course of the river from the high mountains through the great plateaus to the arid plains and desert areas of the Southwest where this river system is the only large source of water. No other stream in the United States passes through such a wide variety of country. It is one of the few rivers of the world that has carved so much of its course through a desert, and the problems of the development of this territory are unique and unprecedented. The great size of the stream cannot be overlooked in a comparison with other rivers; but even more important is the fact that the Colorado serves as the basis upon which the settlement and development of the Southwest have been built.

GEOGRAPHY

The upper continuation of the Colorado River, formerly known as the Grand River,[7] rises in Colorado in the Rocky

[6] Including the Great Lakes, a total length of from 2,100 to 2,200 miles may be computed. See Heilprin and Heilprin, *op. cit.*, p. 1610.

[7] On July 25, 1921, the name of the Grand River was changed to Colorado River.

Mountains at Grand Lake, five or six miles west of Long's Peak.[8] It flows in a southwesterly direction for about 350 miles to its junction with the Green River in southwestern Utah. To this point its principal tributaries are the Gunnison River and the Dolores River, which have their sources in northern and central Colorado. The Green River is the longest tributary of the Colorado. It heads near Fremont's Peak in the Wind River Mountains of southwestern Wyoming, and flows in a southerly direction into Utah for 650 miles to its junction with the Colorado. Its principal tributaries are the Duchesne, Uinta, Price, and San Rafael rivers which rise in Utah and the Yampa and the White rivers which rise in Colorado.

From its junction with the Green, the Colorado River flows toward the southwest into northern Arizona. Here it takes a westerly direction through the Grand Canyon to Nevada, where it becomes the Arizona-Nevada border. It turns toward the south and forms the border between Arizona and California and later the border between Arizona and Mexico (Lower California). From the southwestern tip of Arizona the river is bounded on both sides by Mexican territory, and it empties into the Gulf of California. The chief tributaries of the Colorado River after its junction with the Green are the Fremont River and the Escalante River which rise in Utah and join the Colorado in the same state, the San Juan River which rises in Colorado and joins the river in Utah, the Little Colorado River which rises in New Mexico and joins the Colorado in Arizona, the Paria River and Kanab Creek which rise in Utah and join the Colorado in Arizona, the Virgin River which rises in Utah and joins the Colorado in Nevada, the Williams River which rises in Arizona and joins the Colorado in the same state, and the Gila River which

Before that time, the term "Colorado River" was applied to the stream from its mouth in the Gulf of California to the junction of the Green and the Grand but not beyond. The Grand is more in alignment with the main stream than is the Green, and although it is not as long as the Green its volume is greater. Therefore, the Grand River was officially designated by Congress to be the upper continuation of the Colorado River, and its name was changed accordingly. See *The Statutes at Large of the United States of America*, Vol. XLII, Part I, p. 146. See also House Report No. 1354, 66th Congress, 3d Session, *Change of Name of Grand River to Colorado River* (1921), pp. 1–4.

[8] J. W. Powell, *Canyons of the Colorado* (1895), p. 17.

rises in southwestern Arizona, New Mexico, and Mexico and joins the Colorado in Arizona.[9]

This huge basin of the Colorado River includes parts of seven states—the southwestern part of Wyoming, the western sections of Colorado and New Mexico, all of Arizona except the southeastern corner, the eastern and southern sections of Utah,

TABLE II.—DRAINAGE BASIN AREA BY STREAM BASINS*

	Square Miles	Percentage of Total Area
Green River	44,000	18.0
Upper Colorado (or Grand River)	26,000	11.0
San Juan River	26,000	11.0
Fremont River	4,600	2.0
Paria River	1,400	.5
Escalante River	1,800	.6
Kanab Creek	2,200	.9
Little Colorado River	26,000	11.0
Virgin River	11,000	4.0
Miscellaneous rivers	44,000	18.0
Gila River	57,000	23.0
Total	244,000	100.0

* Senate Document No. 142, 67th Congress, 2d Session, *Problems of the Imperial Valley and Vicinity* (1922), p. 3.

the southeastern part of Nevada, and the southeastern edge of California.[10] This territory is divided naturally into two parts: (1) the plateau region, comprising the upper portion of the area; and (2) the desert country, but little above sea level, comprising the lower portion. The upper basin is bounded on north, east, and west by ranges of mountains attaining an altitude of from 8,000 to 14,000 feet above sea level. On the north and east the Wind River Mountains and the ranges of the Continental Divide are the boundary; on the north and west the Gros Ventre and Wyoming mountains and the Wasatch Range form the rim of the basin. Throughout the winter, these mountains are covered with snow, which melts during the summer months and creates the thousands of streams which combine eventually to form the Colorado River. These streams flow

[9] See Table II.

[10] See Table III.

through mountain passes and valleys out on to the great plateau region. This region begins near the headwaters of the Colorado and Green rivers and extends to the Mogollon Escarpment, a step of from 3,000 to more than 4,000 feet which separates the highlands from the lowlands. The general surface of the plateau region is from 5,000 to 8,000 feet above sea level, but the

TABLE III.—DRAINAGE BASIN AREA BY STATES*

	Square Miles	Percentage of Total Area
Wyoming	19,000	8
Colorado	39,000	16
New Mexico	23,000	9
Utah	40,000	16
Arizona	103,000	42
Nevada	12,000	5
California	6,000	3
Area in United States	242,000	99
Area in Mexico	2,000	1
Total	244,000	100

* Senate Document No. 142, 67th Congress, 2d Session, *op. cit.*, p. 3.

streams have cut deep canyons through it and thus flow at a much lower level.[11] For more than a thousand miles, or about two-thirds of the length of the river, the Colorado has cut a narrow gorge through this tableland, a gorge that in places is over a mile deep. Every tributary throughout this region has cut a canyon, and each river and creek entering the tributaries has cut another. The result is that much of the upper part of the basin is traversed by a labyrinth of deep gorges. The higher plateaus are covered with forests of pine and fir, but the lower levels are dry and treeless. It is in these arid plains that the gigantic canyons have been carved. If the flood storms of the streams of this region were falling on the plains, the adjacent country would be washed and the stream channels would have been little below the general level. Under the prevailing conditions, however, with the source of nearly all of the water in the

[11] Powell, *op. cit.*, p. 28.

high mountains, the canyons are being cut deeper and deeper, while thousands of acre-feet of silt are being carried annually to the lower regions of the river.

The Mogollon Escarpment begins in the Sierra Madre Mountains of New Mexico and extends northwestward across Colorado far into Utah, where it ends at the margin of the Great Basin. Below this great step stretches the lower basin of the Colorado River. This is a country of low, hot, arid plains and valleys, although here and there lone ranges of volcanic mountains rise to an altitude of from 2,000 to 6,000 feet. Rainstorms are infrequent, and there are few springs and creeks. The two great sources of water in this desert area are the trunk of the Colorado and the trunk of the Gila.

For some distance below the mouth of the Virgin River, where the Colorado forms the boundary between Nevada and Arizona and between California and Arizona, the river flows through a number of canyons carved out of volcanic rock. Between these canyons the river has a low but rather narrow flood plain, which supports mesquite and occasional cottonwood groves and which may be cultivated if proper irrigation and protection from floods are secured. This region of irregular mountain chains and valleys continues until Laguna Dam is reached. Toward the west the desert stretches into Nevada and California. Toward the east, across Arizona, the arid plains rise until the pine-covered Sierra Madre Mountains of New Mexico are reached. Here the Gila River finds its chief source of supply, although it descends very quickly to the desert valley below, where frequently its bed is dry during prolonged periods of drought. The Gila region is subject to sudden and violent storms, and on such occasions the water may flow down the river in a great destructive flood, carrying tons of silt and debris to the delta region below.

The apex of the river delta is at Pilot Knob, about seven miles below Yuma, Arizona. From an elevation of about 120 feet near this point the land slopes away toward the south, southwest, and west.[12] On the California side, the desert

[12] House Document No. 504, 62d Congress, 2d Session, *Message of the President of the United States, etc.* (1912), pp. 163, 164.

stretches toward the west, almost to the Pacific Ocean. At one time the Gulf of California extended much farther to the northwest, into territory which is now southeastern California; but the Colorado brought so much silt to the Gulf that eventually its delta became a dam which separated the northern part of the Gulf from the part below. Gradually the upper waters were cut off from the sea and evaporated, and a great valley, largely below sea level, was created. This valley is called the Salton Basin; and within it lies Imperial Valley, one of the richest agricultural areas of the world. During this geological development the Colorado River cut a new channel to the lower Gulf, and thus for centuries it has flowed along the rim of the Salton Basin at a level of from 100 to 300 feet higher than the valley floor.

CHARACTERISTICS OF THE STREAM

Changes in volume of water carried.—The volume of water in the Colorado River may vary widely from year to year, from month to month, and, at times, from day to day. Even the seasonal variations are not accurately predictable. The records of stream flow show that the average annual runoff at Yuma for the period 1902 to 1920 was about 17,300,000 acre-feet.[13] Had the irrigated area in 1922 been under irrigation during that period, the average flow would have been about 16,000,000 acre-feet. The average flow from 1878 to 1922 is estimated to have been between 13,000,000 and 14,000,000 acre-feet; but for periods of more than twenty years the average annual flow may be less than 11,000,000 or more than 16,000,000 acre-feet. Under the 1922 conditions the annual water supply likely to reach Laguna Dam varies from about 6,000,000 acre-feet a year to about 27,000,000, but it is not impossible that these limits may

[13] See the following *Water-Supply Papers,* Department of the Interior, United States Geological Survey:

No.	Page	No.	Page	No.	Page	No.	Page
85	20	249	46	359	11	479	17
100	25	269	44	389	21	509	26
133	32	289	33	409	24	529	19
177	16	309	25	439	23	549	22
213	29	329	24	459	21	569	20

be exceeded in both directions.[14] The average flow at Yuma for the period 1902 to 1935 was about 14,770,000 acre-feet.

By far the greatest contribution to the stream's flow comes from the mountainous upper basin. Even the Gila finds a part of its source in New Mexico. At the junction of the upper Colorado and the Green rivers, the annual discharge of the upper Colorado is 6,940,000 acre-feet and that of the Green is 5,510,000 acre-feet. The upper Colorado, Green, and San Juan rivers together drain 86 per cent of the area of the Colorado River Basin.[15] The greater part of the precipitation in the mountains and high plateaus of the upper basin is in the form of snow, which accumulates to great depths during the winter months.

TABLE IV.—AVERAGE DISCHARGE OF PRINCIPAL TRIBUTARIES*

	Percentage of Total Discharge	Discharge in Acre-Feet	Square Miles	Percentage of Total Area
Upper Colorado (Grand River)	40	6,940,000	26,000	10
Green River	32	5,510,000	44,000	18
San Juan River	14	2,700,000	26,000	10
Other areas except Gila	8	1,560,000	91,000	39
Gila River	6	1,070,000	57,000	23
Total	100	17,780,000	244,000	100

* Senate Document No. 142, 67th Congress, 2d Session, *op. cit.*, p. 2.

This snow melts during the spring and early summer, and sends down the Colorado a flood of water which usually reaches its peak in June. Practically no flow is received from snowfall in the lower basin, although in the northern part of that division there are mountains which rise to an altitude of several thousand feet. The low-water season usually begins in August, and may last from three to seven months. On several occasions the lower Colorado has been practically dry during weeks of extreme drought in the low-water period.

The average annual discharge of the Gila is not great, al-

[14] William Kelly, "The Colorado River Problem," *Transactions of the American Society of Civil Engineers*, Vol. LXXXVIII (1925), Paper No. 1558, p. 328.

[15] See Table IV. Note that the water supply from the various branches is by no means in proportion to the area drained. The discrepancy is due to the wide diversity of climatic and topographic conditions.

though the river drains an area of 57,000 square miles. However, the discharge is extremely variable. In 1916 the discharge of the river at its mouth was 4,500,000 acre-feet, while in some other years the total has been less than 100,000 acre-feet. The river is subject to unpredictable "flash" floods—floods of very short duration but which at times may be even greater than the highest peak floods in the Colorado. On January 16, 1916, for instance, the flow reached the tremendous volume of 220,000 cubic feet per second.[16] At times the waters of the Gila have tended to steady the flow of the Colorado; at other times, however, they have served to accentuate the already wide variations in stream flow.

Quantities of silt carried.—Another important characteristic of the Colorado River is the quantity of silt carried by the stream. For centuries the Colorado and its tributaries have been eroding their beds and banks and carrying thousands of acre-feet of sediment, a part of which is deposited on the alluvial valleys during periods of overflow and a part of which reaches the Gulf of California, where the river is continually extending and enlarging its delta. The Colorado has a silt content three times that of the Ganges and ten times that of the Nile. With the possible exception of the Tigris, it is the greatest silt-bearing stream in the world.[17] The quantity of silt carried at any one time varies widely. In general, the volume varies with the volume of water in the stream, but the variation is not a proportional one. The velocity of the flow, the stage of the flow (rising or falling), the time of the year, and the type of silt present (depending upon its place of origin) also influence the volume and percentage of silt carried.

Numerous estimates have been made of the annual amount of silt carried down by the Colorado. Many of the early estimates were far too low, since only suspended silt was measured and no consideration was given to the heavier matter known as bed silt. For some years the amount of silt being brought down

[16] G. E. P. Smith, *The Colorado River and Arizona's Interest in Its Development* (1922), pp. 532–33.

[17] Senate Hearings before the Committee on Irrigation and Reclamation, 69th Congress, 1st Session, *Colorado River Basin,* Part 1 (1925), p. 27.

the river was estimated at approximately 100,000 acre-feet per year average. The result of careful investigations made by the Department of Agriculture, however, fixes 138,000 acre-feet as the total annual silt load of the river at Yuma, exclusive of the Gila silt;[18] or, it is estimated that a fair average would be 137,000 acre-feet at the lower end of the canyon section of the river, since an additional 1,000 acre-feet of bed silt is believed to be moved from the canyon section past Yuma.[19] This is enough silt to cover approximately 214 square miles of territory to a depth of one foot. With such a tremendous annual silt load, it is easy to understand how the Colorado and its tributaries throughout the centuries could carve their great canyons and build a dam large enough to isolate the Salton Basin from the sea.

Silt has been the creator of much of the agricultural wealth of the lower Colorado River Basin. It contains nitrogen, phosphoric acid, and potash, and thus has an important fertilizing value. Without silt there would be no delta, no fertile alluvial valleys, and no rich farming lands. On the other hand, silt is also the greatest menace to irrigation and water control. When irrigation water containing silt is applied to a field, the silt is deposited near the intake of water to the field. This means that at regular intervals the farmer must move the deposited silt to another part of the field in order to keep the land surface below the intake level. It is estimated that the cost of this silt disposal to farmers in the Imperial Valley is about $2.00 per acre per

18 The silt of the Gila River has been excluded for several reasons. Since Boulder Canyon is far upstream from the mouth of the Gila, its silt could not affect that project. It cannot affect the All-American Canal, since Imperial Dam, the intake point for the All-American Canal, is also north of the Gila. Therefore, when the All-American Canal is completed, Gila silt will no longer find its way into the Imperial Valley canal system. Also, proposed developments on the Gila, if carried through, will eliminate much of the Gila silt which now reaches the Colorado. It should be noted, however, that the Gila carries large quantities of silt, even though it is frequently dry and its water is fairly clear during low flow. During high flow it is an extremely muddy river, and it will continue to pour its silt on the delta and tend to raise the bed of the Colorado until reservoirs are built on its own course above Yuma. The Williams River, another tributary of the Colorado which is also south of Boulder Canyon, has relatively little effect upon the quantity of silt in the river, since it is often dry for long periods of time.

19 Samuel Fortier and Harry F. Blaney, *Silt in the Colorado River and Its Relation to Irrigation* (1928), pp. 61–62.

year. Another silt expense is incurred by the Imperial Irrigation District, which spends about $1.00 per acre per year as the cost of preventing the canal system from becoming clogged with silt. This expense reaches the farmer in the form of higher water charges. Thus in the Imperial Valley there is a fixed annual burden of at least $3.00 per acre, and the total annual burden due solely to silt, according to an estimate made in 1928, is not less than $1,500,000.[20] In addition, all water used for domestic or municipal purposes must be filtered, and this is another expense that must be borne by the valley people. Both the Imperial Irrigation District and the Bureau of Reclamation have spent large sums in building headworks to prevent the heavier silt from entering the Imperial and Yuma irrigation systems. Still it has not been found feasible to prevent the finer sediment from entering the irrigation ditches and finally reaching the fields. It is not improbable that in the future other structures will be built to utilize the waters of the Colorado. If such structures are built, it is to be expected that silt control will be an important and expensive influence upon the building plans and the type of development that will take place.

During medium and high stages of flow in the river, the water is frequently loaded with sand, which is ground to fine silt and carried to the lower basin. Where the river passes through wind-laid material, erosion is especially easy. During flood periods boulders weighing many tons may be rolled slowly down the stream and eventually reduced to silt. From the canyon section to the mouth of the river the slope of fall is not as steep as it is in the upper regions. Thus in the lower basin the heavier silt is transported as bed silt and only that which is very fine in texture remains in suspension. Tests have shown that any velocity of flow that is practical for an irrigation canal will carry in suspension most of the finer silt of the Colorado. Therefore this silt is carried through the irrigation ditches and deposited eventually upon the farm lands. There it acts much like Portland cement. It seals up the relatively porous soil and

[20] House Hearing before the Committee on Irrigation and Reclamation, 70th Congress, 1st Session (1928), *Protection and Development of the Lower Colorado River*, p. 481.

impairs its texture and productivity. It acts as a more or less impervious blanket, appreciably lessening the penetration of water into the soil. In the Imperial Valley, waste and drainage waters immediately precipitate their silt load when they come into contact with the still and salty waters of the Salton Sea. This precipitation forms deltas which obstruct the natural flow and cause the water to back up and flood the farm lands near by. In 1924 the Imperial district spent about $31,000 in dredging the New River outlet to permit the flow of the waste waters to continue.

The deposition of so much silt in the delta region has resulted in an elevation of the river bed and has thus caused a serious flood menace to the Imperial Valley below the bed of the river. For centuries the Colorado has meandered over its delta without human interference. If its channel became too high, it formed a new channel at a lower level and used it until the accumulations of silt clogged its bed again. This tendency of the Colorado to change its course was strongly resisted by nature. The annual inundations resulted in a dense growth of vegetation in the fertile soil of the delta and thus imposed an obstacle to the erosion of pronounced channels away from the river bed. Yet for more than fifty years the Colorado has been ready to leave the ridge on which it flows and to drop into the Imperial Valley. Late in the nineteenth century, when white men began to cultivate the rich delta lands, levees were built to hold the river in one course. However, in June 1905 the Colorado turned its entire flow into the Imperial Valley, and the break was not closed until February 1907, after an expenditure of more than $2,000,000. In 1909 the river turned into the Volcano Lake Basin, which served as an outlet to the Gulf until 1918, when another change in direction of the river channel occurred. The Pescadero Basin, into which the river has flowed since 1922, has filled so rapidly that a tendency to move off the debris cone at the end of the diversion channel is quite manifest. Some $8,000,000 have been spent in building levees along the lower reaches of the Colorado to attempt to confine the river to its course to the Gulf.[21] The river destroyed many of these

[21] Fortier and Blaney, *op. cit.*, pp. 28–29.

levees and continued to build up its bed, while the building of higher and higher levees became increasingly difficult and expensive. As time went on it became more and more apparent that unless the volume of flow could be regulated and the channel freed from the deposition of silt, the Imperial Valley and its property values of over $100,000,000 were doomed.

HISTORY OF DEVELOPMENT OF THE LOWER RIVER

NAVIGATION

Long before the silt and flood dangers of the Colorado River were known, the stream had begun to play an important part in the exploration and development of the Southwest. The American pioneers who settled in the Imperial Valley and diverted water from the river to irrigate the desert lands were not the first white men to reach the lower Colorado region and to traverse its vast deserts. Before the middle of the sixteenth century the Spanish conquerors of Mexico had pushed northward by land and by sea in a fruitless search for the cities of gold which supposedly existed somewhere in this unexplored territory. On July 8, 1539, Francisco de Ulloa sailed from Acapulco, Mexico, with a fleet of three vessels, and shortly thereafter reached the head of the Gulf of California.[22] He did not see the Colorado River, but the turbid condition of the water at the head of the Gulf convinced him that there was a great river near by. In 1540, Hernando de Alarcón sailed to the head of the Gulf of California, discovered the mouth of the Colorado, and sailed up the stream for over 200 miles. He was the first European navigator of the Colorado River. Other explorers who reached the Colorado during the sixteenth century traveled mainly by the overland routes. During the seventeenth and eighteenth centuries a number of Spanish padres explored the Colorado region, and attempted to establish missions there. Early in the nineteenth century American trappers appeared on the river, and soon afterward many American travelers selected the lower Colorado route for their journeys to and from the West Coast.

[22] House Document No. 79, 57th Congress, 2d Session, *First Annual Report of the Reclamation Service* (1902), pp. 121–23.

From 1846 until the beginning of the Civil War, the lower Colorado was visited and surveyed by exploring parties under the direction of the War Department. By the terms of the treaty with Mexico signed at Guadalupe Hidalgo on February 2, 1848, the boundaries of the United States had been extended to include most of the lower basin of the Colorado. The treaty had also provided that the vessels of the United States were to have free passage of the Gulf of California and the Colorado River, and that neither country without consent of the other would construct works on the Colorado or Gila rivers which would obstruct navigation.[23] Subsequently the Gadsden Purchase Treaty of December 30, 1853, extended the American border still farther but canceled the navigation articles of the previous treaty. However, the United States was still guaranteed the free and uninterrupted passage of vessels and citizens on the Gulf of California and the Colorado River.[24]

In 1850–51 a reconnaissance of the Gulf of California and the Colorado was made under the direction of Lieutenant George H. Derby of the Topographical Engineers. The purpose of the expedition was to find a route for transportation by water of supplies to Fort Yuma. The trip was made in the schooner "Invincible," which drew too much water to proceed beyond the mouth of the river. It was concluded that the Colorado was not navigable to ocean boats,[25] but river steamers could be used, and were used, in spite of the shallowness of the stream, the rapids, and the shifting channel. In 1830 a stern-wheeler named "Yuma" came up the river to Fort Yuma, and in 1851 similar trips were made by the "Uncle Sam."[26] In 1857 Lieutenant J. C. Ives navigated the Colorado from its mouth to Las Vegas Wash in a metal steamboat named "Explorer," which had been built in Philadelphia and shipped in sections to the mouth of the Colorado via San Francisco.[27] In 1866 Captain

[23] *United States Statutes at Large and Treaties from December 1, 1845, to March 3, 1851*, Vol. IX, pp. 926, 928–29.

[24] *United States Statutes at Large and Treaties from December 1, 1851, to March 3, 1855*, Vol. X, p. 1034.

[25] Senate Executive Document No. 81, 32d Congress, 1st Session, *Report of the Secretary of War* (1852), pp. 2–22.

[26] E. C. La Rue, *Colorado River and Its Utilization* (1916), pp. 19, 20.

[27] Lieut. J. C. Ives, *Report on Colorado River of the West*, p. 86.

Thomas E. Trueworthy in the steamboat "Esmeralda" went up the Colorado as far as Callville, near the mouth of the Virgin River, several miles above the highest point attained by Ives;[28] but the chief river traffic, which developed about 1860 and lasted until about 1880, was confined to the section of the river between the Gulf and El Dorado, a mining town some 344 miles above Yuma.

Before the railway reached the Colorado River at Yuma in 1877, all supplies sent from San Francisco to the interior of Arizona traveled by ocean steamer to the head of the Gulf of California, and were there transferred to the river boats of the Colorado Steam Navigation Company to be left at various points along the river and freighted overland to their destinations.[29] Most of the traffic was between the Gulf and Yuma, where overland transportation started for Tucson. Above Yuma the most important port was Ehrenberg, where goods and passengers for Prescott were landed. Farther up the river the development was small, the traffic was very light, and obstructions in the river made navigation so difficult that the business was not worth while. As soon as the Southern Pacific Railroad reached Yuma, the most profitable part of the river business disappeared. The railroad took over the steamship company, and all traffic from the Gulf to Yuma was discontinued. Traffic above Yuma continued for two or three years longer; but by 1883, when the Santa Fe Railroad crossed the Colorado at Needles, transportation by river steamers on the lower Colorado was nearly at an end.

Plans for navigation of the Colorado River were not abandoned immediately, however, with the advent of the railroads. The belief persisted that, with some improvement of the stream bed and banks, transportation on the Colorado could be made to pay. Several years before, in 1866, Colonel James F. Rushing, Inspector of the Quartermaster's Department, Camp Doug-

[28] House Executive Document No. 166, 42d Congress, 2d Session, *Freight to Salt Lake City by the Colorado River* (1872), p. 2.

[29] House Executive Document No. 1, 46th Congress, 2d Session, Part 2, Vol. II, *Annual Report of the Chief of Engineers to the Secretary of War* (1879), pp. 1774–76, 1778–79.

las, Utah, reported that the river was navigable far above Callville and that steamers could sail easily to a point not over 350 miles from Salt Lake City. From this point merchandise for Salt Lake City would have to be taken overland. He intimated that the tales concerning the unsuitability of the river for navigation were circulated by the Colorado Steam Navigation Company and other parties who wished to maintain the monopolies they had established.[30] In 1878 Congress passed an act directing the Secretary of War to survey the Colorado River from Yuma to El Dorado and to estimate the cost of the needed improvements.[31] In 1884 Congress appropriated $25,000 for the improvement of the Colorado River above Yuma,[32] but the money available was too small to permit any great improvements. The investigations showed that the possible traffic development was not worth the tremendous sums necessary to improve the river for navigation. In 1890 it was reported by W. H. H. Benyaurd, Lieutenant Colonel, Corps of Engineers, that there were two small steamers on the Colorado which carried supplies occasionally to mining camps along the river; but again it was concluded that the traffic was so small that the river was not worthy of improvement.[33]

With the decline in importance of navigation on the Colorado River, the rise in importance of water for irrigation began. Naturally there was a conflict in interests, since the greater the diversions, the poorer the navigability of the stream, especially during the periods of low flow. Irrigation interests seemed to be gaining the upper hand; but in 1903 the War Department held that the Colorado River was a navigable stream and that diversions of water which would interfere with navigation were unlawful. Because of this decision, the California Development Company, which was promoting the development of Imperial Valley, sought action by Congress in 1904 to confirm its appro-

30 House Executive Document No. 166, 42d Congress, 2d Session, *op. cit.*, pp. 1–2, 6–8.

31 *United States Statutes at Large,* Vol. XX, pp. 160–62.

32 *United States Statutes at Large,* Vol. XXIII, p. 144.

33 House Executive Document No. 18, 51st Congress, 2d Session, *Colorado River, Arizona* (1890), pp. 2–3.

priation of water. A bill was introduced to have the waters of the Colorado declared more valuable for irrigation than for navigation; but it failed to pass.[34] Then the California Development Company, through its Mexican subsidiary, applied for and received in 1904 a concession from the Mexican government to divert water from the Colorado River in Mexico to the Imperial Valley; but it was stipulated that the diversion should be without injury to the navigability of the river. Yet at times of extreme low flow all of the water in the lower river has been diverted into the Imperial Canal and no formal objections have been made. Soon afterward the United States government began the construction of a diversion dam at Laguna, about ten miles north of Yuma. No provision was made for navigation, even though frequently so much water was taken for the Yuma project that navigation below Yuma became impossible. Therefore, it was apparent that although the Colorado was still legally, technically, and by treaty a navigable stream, navigation on the lower Colorado was no longer of importance. It was realized that river regulation and flood control would benefit both navigation and irrigation but that all plans to improve the river for navigation alone had faded from the picture.

AGRICULTURAL DEVELOPMENT

Long before the arrival of the Spanish explorers the Mohave, Chemehuevis, Yuma, and Cocopah Indians cultivated the overflow lands of the lower Colorado River. Their method was simple. They scattered seeds in the mud as the water receded and left the rest of the work to be done by nature. The cliff dwellers were the first to develop an artificial conveyance of water from the Colorado for agricultural use. They dug ditches in the bench lands where they raised their crops below their cliff houses. The Jesuit padres, who were among the first white men to explore the region, built missions near the river and diverted water to irrigate the surrounding territory. The Mormons, coming from the Great Salt Lake Basin down the tribu-

[34] C. E. Tait, *Irrigation in Imperial Valley, California; Its Problems and Possibilities*, Senate Document No. 246, 60th Congress, 1st Session (1908), pp. 12–13.

taries of the Colorado, diverted water to cultivate the river-bottom lands. Later they built reservoirs, and higher and more extensive areas of land were cultivated.[35] Yet during these early days, no extensive plan of agricultural development was put into operation. Travelers noticed the richness of the soil, but control of the river was considered impossible.

Imperial Valley.—The first important agricultural project in the lower Colorado region to receive serious attention was the Imperial Valley, then known as the New River country, in the Salton Basin. The basin extends from a few miles north of Indio, California, toward the south to its rim some fifteen or twenty miles across the Mexican border. It is about 110 miles long and 40 miles wide, and contains over 1,000,000 acres. The northern, narrow end is called the Coachella Valley, and the southern portion is the Imperial Valley.[36] The total area of the Imperial district in the United States is about 605,000 acres, 515,000 of which are considered irrigable.[37]

It was recognized that the Colorado River could be diverted to the basin, and the original proposal to create an inland sea in the basin to temper climatic conditions gave way to a plan to supply the rich Imperial Valley lands with water for irrigation purposes. The idea of bringing water from the Colorado to the arid lands of the Salton Sink was conceived by Dr. O. M. Wozencraft, who had traveled through that region in 1849. His plan was worked out in collaboration with Mr. Ebenezer Hadley, surveyor of San Diego County, California, and it was proposed to divert water from the Colorado to the valley through the Alamo, an old overflow channel of the river, substantially the plan that was followed years later. Through the efforts of Dr. Wozencraft and his associates the legislature of the state of California passed a resolution in 1859 calling upon Congress to cede to the state for reclamation purposes some 3,000,000 acres of land, including all of the territory now known as the

[35] Lewis R. Freeman, *The Colorado River, Yesterday, Today and Tomorrow* (1923), pp. 371–72; see also La Rue, *op. cit.*, pp. 113–14.

[36] Senate Document No. 246, 60th Congress, 1st Session, *Irrigation in Imperial Valley, California* (1908), p. 5.

[37] See Table V and Figure 2.

Imperial Valley.[38] The proposed act for cession met with the approval of the Public Lands Committee of the House of Representatives in 1862, but at that time Congress was absorbed in the progress of the Civil War and the proposed legislation failed to pass. Until his death Dr. Wozencraft fought for the passage of his bill, but he was never successful. Yet one thing had been established definitely. The surveys had proved that water could be diverted from the Colorado River to irrigate the Imperial Valley.

In 1875–76 a government survey was made by Lieutenant Eric Bergland, Corps of Engineers, United States Army, to determine whether or not it was practical to build a canal entirely within American territory from the Colorado River to the Imperial Valley.[39] The report was unfavorable. It pointed out that deep cuts or tunnels would have to be dug through a sand-hill area and that the cost of such a canal would be prohibitive. However, it was shown that there was a practical route through Mexican territory along some of the old overflow channels, and the effort to get water to Imperial Valley continued. Finally in 1892 Mr. C. R. Rockwood, an engineer, investigated the Alamo Channel route through Mexico to the Salton Sink. On the basis of this report the Colorado River Irrigation Company was organized and Rockwood was sent back to make another survey. The company was a result of a landboom idea, was hastily organized, and was not properly financed. Enough stock was sold to finance the Rockwood survey, but no funds were available to carry on the work. The company failed; and the surveys, which were its most valuable asset, were acquired by a new corporation, the California Development Company, organized on April 24, 1896, under the laws of New Jersey.[40]

Between 1896 and 1902 the Imperial canal system was designed, and construction was begun. The original canal headed at Hanlon's Crossing in California, about 500 feet north of

[38] *The Statutes of California,* passed at the Tenth Session of the Legislature (1859), pp. 392–93.

[39] House Executive Document No. 1, Part 2, 44th Congress, 2d Session, Vol. II, Part 3, *Appendix to Report of the Chief of Engineers* (1876), pp. 337–38.

[40] Senate Document No. 142, 67th Congress, 2d Session, *Problems of Imperial Valley and Vicinity* (1922), p. 71.

TABLE V.—LOWER BASIN—ACREAGE IRRIGATED AND IRRIGABLE IN THE FUTURE*

	Irrigated 1920 Gravity	Future Additional Possible		Total Ultimate
		Gravity	Pump	
United States:				
Above Laguna Dam				
Cottonwood Island	....	1,000	3,000	4,000
Mohave Valley	....	24,000	3,000	27,000
Chemehuevis Valley	....	4,000	....	4,000
Parker project	4,000	100,000	6,000	110,000
Palo Verde Valley	35,000	43,000	....	78,000
Palo Verde Mesa	....	....	18,000	18,000
Chucawalla Valley	....	....	44,000	44,000
Cibola Valley	....	16,000	....	16,000
Isolated tracts	....	1,000	3,000	4,000
Total above Laguna Dam	39,000	189,000	77,000	305,000
Below Laguna Dam				
Yuma project	54,000	15,000	61,000	130,000
Imperial District	415,000	100,000	....	515,000
Imperial extensions				
East Mesa (Coachella Valley)	....	124,000	36,000	160,000
Dos Palmas (Coachella Valley)	....	5,000	....	5,000
Coachella Valley (Coachella Valley)	....	72,000	....	72,000
West Side (Coachella Valley)	....	10,000	23,000	33,000
Total below Laguna Dam	469,000	326,000	120,000	915,000
Total, United States	508,000	515,000	197,000	1,220,000
Mexico:				
Under Imperial Canal	190,000	65,000	....	255,000
Under All-American Canal	....	22,000	8,000	30,000
Delta south of Volcano Lake and Bee River	....	250,000	....	250,000
Sonora	....	210,000	55,000	265,000
Total, Mexico	190,000	547,000	63,000	800,000
Grand total, Lower Basin	698,000	1,062,000	260,000	2,020,000

* Senate Document No. 142, 67th Congress, 2d Session, *op. cit.*, p. 32.

the Mexican border. Here a wooden head gate, known as the Chaffey Headgate, was built. The canal crossed the border, and

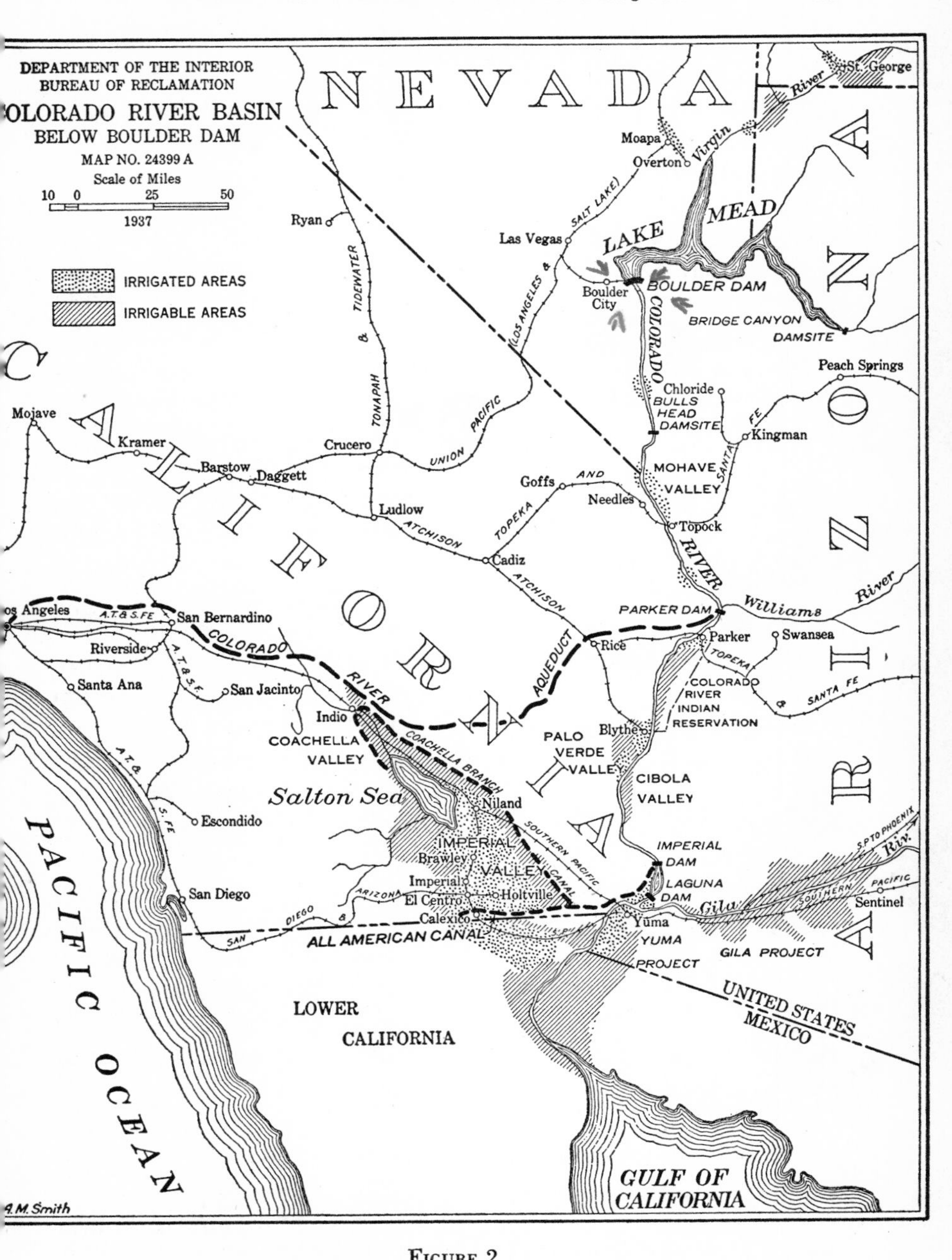
DEPARTMENT OF THE INTERIOR
BUREAU OF RECLAMATION
COLORADO RIVER BASIN
BELOW BOULDER DAM
MAP NO. 24399 A
Scale of Miles
10 0 25 50
1937
IRRIGATED AREAS
IRRIGABLE AREAS
NEVADA
ARIZONA
CALIFORNIA
PACIFIC OCEAN
LOWER CALIFORNIA
GULF OF CALIFORNIA
UNITED STATES
MEXICO
St. George
Virgin River
Moapa
Overton
LAKE MEAD
Las Vegas
Boulder City
BOULDER DAM
BRIDGE CANYON DAMSITE
Peach Springs
Ryan
Chloride
BULLS HEAD DAMSITE
Kingman
MOHAVE VALLEY
Needles
Topock
Goffs
Crucero
Mojave
Kramer
Barstow
Daggett
Ludlow
Cadiz
COLORADO RIVER
Williams River
PARKER DAM
Parker
Swansea
Rice
COLORADO RIVER INDIAN RESERVATION
Blythe
PALO VERDE VALLEY
CIBOLA VALLEY
Los Angeles
San Bernardino
Riverside
Santa Ana
San Jacinto
Indio
COLORADO RIVER AQUEDUCT
COACHELLA VALLEY
COACHELLA BRANCH
Salton Sea
Niland
IMPERIAL VALLEY
Brawley
Imperial
El Centro
Holtville
Calexico
Escondido
San Diego
ALL AMERICAN CANAL
CANAL
IMPERIAL DAM
LAGUNA DAM
Yuma
YUMA PROJECT
Gila
GILA PROJECT
Sentinel
S.P. TO PHOENIX
SALT LAKE
LOS ANGELES &
TONAPAH & TIDEWATER
UNION PACIFIC
ATCHISON TOPEKA AND SANTA FE
A.T.&S.FE
SOUTHERN PACIFIC
SAN DIEGO & ARIZONA
A.M. Smith

FIGURE 2

extended along the Alamo channel through Mexico until it turned north at Sharp's Heading, where the water was carried back into the United States through several smaller canals. Some water was delivered to Mexican lands in 1901, and by 1902 water reached the lands of some 2,000 people who had settled in Imperial Valley. In May 1904 the California Development Company, through its Mexican subsidiary known as La Sociedad de Riegos y Terrenos de la Baja California, was granted the right by the Mexican government to divert 10,000 cubic feet of water per second from the Colorado below the international boundary. Two dredger cuts were made. The first, known as the upper Mexican heading, was just below the border; and the second, known as the lower Mexican heading, was about four miles below. It was through this second cut that the floods of 1905 and 1906 swept across Imperial Valley and into the Salton Sink. The break was not closed until 1907, and was closed then only through the efforts of the Southern Pacific Company, which had loaned money and equipment to stop the flood.

A reorganization of the California Development Company in 1905 had placed the control of its affairs into the hands of the Southern Pacific Company, since the land company had borrowed some $200,000 from the Southern Pacific in June of that year. As a result of this loan and the large flood-control burden that followed, the California Development Company was forced into receivership in December 1909. Yet in spite of these financial difficulties and the ever present flood hazard, the agricultural development of the valley proceeded rapidly. By 1906, 130,000 acres were under cultivation and water rights had been purchased through mutual water companies for over 200,000 acres. By 1910 the irrigated area exceeded 180,000 acres, exclusive of about 15,000 acres in Mexico which were also being supplied by the Imperial Canal. In 1911 the settlers of the valley voted to form an irrigation district. The Southern Pacific Company had acquired the entire system of the California Development Company at the receiver's sale, and the irrigation district voted a bond issue of $3,500,000 to buy out the Southern Pacific's interest for $3,000,000 and to make improvements of $500,000. In June 1916 the sale was consummated when the Southern Pacific Company accepted $3,000,000 of the district's 5 per cent bonds for the property. This purchase included the assets of a new subsidiary Mexican company, the Compania de Terrenos y Aguas de la Baja California, S. A., to which the works of the Imperial canal system in Mexico originally owned by the California Development Company had passed.[41]

Under this new arrangement, the development of the Imperial Valley proceeded at a rapid rate. By 1927 property values had risen to nearly $140,000,000. Population in 1928 had increased to over 75,000 and the annual value of the crops was estimated at $40,000,000. Towns grew and more and more land was brought into production. To maintain this development, the Imperial Irrigation District operated an extensive system of irrigation and protective works. Rockwood Heading was built in 1917 at a cost of $250,000, and a canal 6,000 feet long was dug from Rockwood Intake to Hanlon Heading. The main canal was improved, and lateral canals were constructed, those in the United States totaling more than 2,441 miles. Seventy-three and five-tenths miles of levees were built in Mexico, 47.5 of which were provided with standard railways, and the Pescadero Cut was dredged as another protective measure.[42] Undoubtedly this development would continue if the river could be controlled; but, if the Colorado floods were not subdued, it was only a question of time until the wealth originally created by the diversions from the river would be destroyed by that same river.

The right of the Imperial Irrigation District to divert water from the Colorado River is based on appropriation and use. The first filing was made on May 16, 1895, for 10,000 cubic feet per second on behalf of the proposed Imperial Canal. Similar filings were made during that year and during the four years that followed.[43] Earlier filings had been made for the Blythe Rancho, to which the Palo Verde Irrigation District succeeded, and for the Yuma project, to which the United States Bureau of

[41] Frank Adams, *Irrigation Districts in California* (1929), pp. 336–37, 340–41.

[42] House Document No. 359, 71st Congress, 2d Session, *op. cit.*, pp. 20, 21, 116.

[43] Senate Document No. 142, 67th Congress, 2d Session, *op. cit.*, p. 74.

Reclamation is the successor. Later filings were made for the Yuma and Colorado River Indian reservations and for the Parker Indian project. Filings of still later dates were made by upper basin states both for power and for irrigation. In 1904 the California Development Company appealed to Congress for recognition of the claim to water for irrigation from the Colorado River. Nominally the Colorado River was a navigable stream, and there were, in fact, steamboats on the river. Congress did not recognize the diversion rights of the company, and refused to have the river declared non-navigable. At about this same time, the California Development Company, through its Mexican subsidiary, was granted the right to divert 10,000 cubic feet per second from the river in Mexican territory. The Imperial Irrigation District succeeded to all of these rights of the California Development Company. The rights, made under the old code filings, have never been seriously questioned. The Imperial Canal has a diversion capacity in excess of 7,000 cubic feet per second. From 1922 to 1927 the annual diversions by the entire canal system were about 3,000,000 acre-feet.[44]

All operations of the California Development Company in Mexico, and later of the Imperial Irrigation District, had to be carried on through a Mexican subsidiary, La Sociedad de Riegos y Terrenos de la Baja California, later changed to Compania de Terrenos y Aguas de la Baja California, S. A. There had been no agreement entered into between the United States and Mexico under which Mexico granted to the United States or any of its citizens the right to conduct water across Mexican territory. Neither had there been a direct grant of any such right to an American corporation. The concession was made to a Mexican corporation, whose operations, so far as constructing works and conducting water are concerned, are confined entirely to Mexico. According to Mexican law, the stock of the Mexican companies could not be owned by the American corporations. The stock is held by the individual members of the board of directors of the Imperial Irrigation District (formerly by the individual members of the board of directors of the California Development Company), and each director on leaving the board assigns

[44] Adams, *op. cit.*, pp. 339–40.

his interest to his successor. A complete staff of officials must be kept in Mexico, all business with the Mexican company must be transacted in Spanish, and the metric system must be used in all operations in Mexico.[45]

The original Mexican concession of May 17, 1904, provided that water could be diverted from canals running through Mexican territory to be used in irrigating lands in Lower California. The amount of water to be diverted, however, was limited to one-half of the volume passing through the canals. The area irrigated from the Imperial canal system in the United States greatly exceeds the area irrigated from that system in Mexico, and the amount of water needed in Mexico is much less than half of the total.[46] Thus Mexico takes half of the water only during periods of shortage when the American lands may suffer from drought because of the requirements of the concession. Although it is usually necessary to construct a temporary weir across the Colorado River during certain low-water periods of the summer to facilitate diversion, there have been only a few brief periods of shortage, such as from August 4 to October 16, 1924, when the entire flow of the river was diverted by the Imperial Canal and a minimum flow of 1,295 cubic feet per second was reached on September 11.[47] Still, during these periods of shortage, the American farmers have been forced to decrease the irrigation of their lands so that Mexican lands could be irrigated fully. This condition existed in spite of the fact that the canal was built and supported largely by American interests. In addition the price of water was raised from time to time on the American side until in 1925 it amounted to $1.00 per acre-foot plus a tax rate of $2.50 per acre, while on the Mexican side only one raise was permitted, from 50 cents to 85 cents an acre-foot. The American farmers did not believe that the Mexican farmers were paying a just proportion of the cost of operating the system.[48] Their resentment was heightened during pe-

[45] Senate Document No. 103, 65th Congress, 1st Session, *Colorado River in Its Relation to Imperial Valley* (1917), pp. 5, 15, 16.

[46] See Table V.

[47] Adams, *op. cit.*, pp. 339, 342.

[48] Senate Hearings before the Committee on Irrigation and Reclamation, 68th Congress, 2d Session, *Colorado River Basin* (1924), p. 170.

riods of flood danger when the Mexican government impeded flood control work although levees built in Mexico by American capital protected Mexican lands as well as American lands. Duties had to be paid on equipment and supplies sent across the border. Inefficient Mexican labor had to be used. Valuable time was wasted in securing permission to carry on the work from a seemingly disinterested government. Revolutionists and bandits raided construction camps, and even threatened to destroy protective works. Under these circumstances, it is not surprising that comparatively early in the development of the Imperial Valley the American farmers began to agitate for protective works and a canal located entirely within American territory.

The agitation for an all-American canal began to crystallize into definite action in 1917. A contract was executed between the Secretary of the Interior and the Imperial Laguna Canal Company to irrigate a tract of land comprising 120,000 to 200,000 acres adjoining the Imperial Valley by means of an all-American canal some thirty miles in length.[49] The water was to be taken from the Colorado River at Laguna Dam, and surveys of the route were begun. Later it was proposed to construct the canal to irrigate the entire Imperial Valley. On February 16, 1918, a co-operative agreement was entered into between the Imperial Irrigation District and the Department of the Interior to make a complete investigation, survey, and cost estimate of an all-American canal from Laguna Dam to Imperial Valley. Two-thirds of the cost of this investigation, or $30,000, was to be borne by the district and one-third, or $15,000, by the government. The investigations were to follow the general plan agreed upon by the All-American Canal Board, composed of Mr. C. E. Grunsky, representing the Imperial Irrigation District, Mr. Elwood Mead, acting for the University of California, and Mr. W. W. Schlecht of the Reclamation Service.[50] In October 1918 the All-American Canal Association of Los Angeles was organized to promote the construction of the canal to serve

[49] *Engineering News-Record,* Vol. LXXVIII, No. 13 (April 2, 1917), p. 235.

[50] *Seventeenth Annual Report of the Reclamation Service* (1918), p. 382.

lands then unproductive because they lay above the Imperial Valley canal system. The proposed canal was to head at Laguna Dam and to follow the main canal of the Yuma project for a distance of ten miles to Siphon Drop, about four miles north of the town of Yuma. From Siphon Drop the canal would run in a southwesterly direction past Pilot Knob and the old heading of the Imperial Canal to within 400 feet of the international boundary line, which it would parallel for four miles, well into the sand-hill area. There it would turn northwest through the sand hills and would pass north of Calexico and across New River. It would terminate in the main canal now constructed on the west side of Imperial Valley. The total length of the proposed line was a little over 76 miles.[51]

On October 23, 1918, an agreement was entered into between the Secretary of the Interior and the Imperial Irrigation District to provide for the extension of the Imperial Canal to Laguna Dam. Such an extension would move the intake of the Imperial Canal to Laguna Dam and would avoid the construction of further temporary dams in the river at Hanlon Heading, the intake then in use. The district pledged itself to pay the United States $1,600,000 in twenty installments, beginning December 31, 1919, for the privilege of using the dam. Immediately upon the execution of the contract the district was to make surveys and secure information concerning the specifications and costs of an all-American canal. The district agreed to build the canal at as early a date as possible and within a reasonable time. The agreement was ratified by the voters of the Imperial Irrigation District on January 21, 1919, even though the All-American Canal Board report had not as yet been made. When the report was made, later in 1919, Secretary Lane of the Department of the Interior decided that the project was impractical and released the district from the obligation to build the canal.[52]

In 1920 another survey and general investigation of the proposed canal was begun by the United States Reclamation Service. Local interests raised $40,000 to add to a federal

[51] *Eighteenth Annual Report of the Reclamation Service* (1919), pp. 384–85.

[52] William Kelly, "The Colorado River Problem," *Transactions of the American Society of Civil Engineers*, Vol. LXXXVIII, Paper No. 1558 (1925), p. 352.

appropriation of $20,000 to defray the costs of this survey,[53] but construction was not begun. The Imperial District made the payments in accordance with the agreement of October 23, 1918, but financial difficulties and the possibility that the United States might undertake the project delayed further action. The bills proposed in Congress for this project failed to pass. Eventually, however, the all-American canal idea became identified with the larger problem of river control and regulation. Its supporters achieved their goal when it became a part of the Boulder Canyon Project Act.

Mexican lands.—The Mexican portion of the Imperial Valley lies south of and adjacent to the Imperial District in California. It extends from east to west for about 50 miles along the border and to the south for a maximum distance of about 20 miles. According to the Mexican concession of 1904, Mexican lands were entitled to one-half of the water running through the Imperial Canal. Under this agreement irrigation began on the Mexican side of the border in 1906 but developed very slowly until 1916. Since that time development has been at a much more rapid rate. In 1915, 40,000 acres were being irrigated in Mexico; in 1918, 118,530 acres; in 1920, 190,000 acres; and in 1925, 217,000 acres.[54] An estimated total of 800,000 acres may be irrigated in Mexico from the waters of the Colorado River, and of this total 255,000 acres may be supplied from the present Imperial Canal.[55] These lands are held in large tracts, chiefly by American interests through Mexican subsidiaries. It is feared that there is not enough water to irrigate all of the irrigable lands in the Colorado region and the use of water in Mexico might serve to limit the expansion in the United States. Thus the rapid development of the Mexican lands has proved to be a disturbing factor on the American side of the border. To solve this and similar problems an act was passed by the United States Congress on March 3, 1927, which

[53] *Engineering News-Record*, Vol. LXXXV, No. 12 (September 23, 1920), p. 626.

[54] Senate Document No. 142, 67th Congress, 2d Session, *op. cit.*, pp. 6, 71. See also House Report No. 1657, 69th Congress, 2d Session, *Boulder Canyon Reclamation Project* (1926), p. 16.

[55] See Table V and Figure 2.

authorized the President to appoint three special commissioners to co-operate with representatives of Mexico in a study regarding equitable use of the waters of the lower Rio Grande, the lower Colorado, and the Tia Juana rivers.[56] The representatives of the two governments, called the International Water Commission of the United States and Mexico, met on three different occasions but finally adjourned on November 9, 1929, when they were unable to reconcile their differences.[57]

Coachella Valley.—The Coachella Valley occupies the northern portion of the Salton Basin. The water supply comes from the high mountains down the Whitewater River and its tributaries. These waters sink into an underground water basin and are recovered for irrigation purposes through the use of artesian wells. In 1894 the first artesian well was bored in the valley near Indio. New methods of drilling were invented, and by 1900 it cost comparatively little to drill an artesian well in this district.[58] More land was brought into production, and by 1918 it was realized that the annual withdrawals from the artesian basin were greater than the annual replenishments. To study this situation the Coachella Valley County Water District was organized with a gross area of 992,320 acres, including the surrounding territory. The engineers of the district reported that there was sufficient water in the artesian basin to cultivate only 10,000 acres, although there were over 260,000 acres of irrigable land in the district.[59] In 1919 the district constructed a large dam across the Whitewater River to spread the flood waters and thus to permit them to sink more easily into the pervious gravels of the artesian basin; but by 1929 some 16,000 acres were being cultivated, thus leading to a large overdraft which the mountain rains could not replenish. The great hope of the Coachella Valley is the construction of the All-American Canal with a branch circling the rim of the valley. With such a canal it is estimated that 108,000 acres of Coachella lands can be irrigated by gravity

[56] *United States Statutes at Large,* Vol. XLIV, Part 2, p. 1403.

[57] House Document No. 359, 71st Congress, 2d Session, *op. cit.,* pp. 11, 161–62.

[58] Senate Hearings before the Committee on Irrigation and Reclamation, 69th Congress, 1st Session, *Colorado River Basin* (1925), Part 2, pp. 53–54.

[59] Adams, *op. cit.,* pp. 374–75.

and that in addition some 30,000 acres can be irrigated with pump lifts of not over 250 feet.[60]

Yuma project.—The Yuma project is located in Yuma County, Arizona, and in Imperial County, California; but much the greater part of the project is located on the Arizona side of the Colorado River. On the California side the project comprises the lands of the Yuma Indian Reservation and occupies that part of the Yuma Valley which extends for about ten miles northeast from Yuma to Laguna Dam. On the Arizona side the project extends from Yuma to the Mexican border, a distance of about seventeen miles. The valley is about six miles wide.[61]

During the period from 1890 to 1899 settlement on the lands of the Yuma Valley was begun and irrigation systems were developed. The early settlers faced many difficulties. There were contests for land titles against claimants of a Spanish grant; there were floods from the Colorado River; and the irrigation systems were poorly constructed and poorly planned. To receive aid in solving these problems the settlers appealed to the Reclamation Service. Investigations for the development of irrigation projects along the Colorado River were begun in October 1902. A survey was made of the Colorado River Valley from 100 miles above Needles to the Mexican border, a distance of 400 miles. It was decided that the Yuma project was the most promising for immediate development, and extensive preliminary surveys were made during the winter of 1903–4.[62] In November 1903 the Yuma County Water Users' Association was organized, and practically all of the private lands within the limits of the Yuma project were subscribed and placed under contract to take water in accordance with the provisions of the Reclamation Act.[63] On April 8, 1904, a board of engineers

[60] Colorado River Commission of the State of California, *Colorado River and Boulder Canyon Project* (1930), pp. 247–48.

[61] Senate Document No. 142, 67th Congress, 2d Session, *op. cit.*, p. 62.

[62] *Ninth Annual Report of the Reclamation Service* (1910), pp. 73–74.

[63] For many years the United States government had been content to leave the reclamation of arid lands to individual effort and acts were passed to encourage private enterprise. Toward the turn of the century it became evident that more thorough treatment was necessary and that it could be secured only through national action. The government was not willing to part with its ownership of lands in the arid region as a whole or in large compact areas, and without ownership of the lands the con-

approved the plans for the project, and on May 10, 1904, the Secretary of the Interior authorized the project and set aside $3,000,000 from the reclamation fund for the work.[64] The right to divert water from the Colorado River (considered a navigable stream) for the Yuma project was derived from an act of Congress approved April 21 of that year,[65] and on July 8, 1905, an appropriation for 3,000 second-feet of water for use in Yuma County, Arizona, was filed at the Yuma County recorder's office.

The investigation had shown that bedrock conditions made it impossible to build a high diversion dam at the northern end of the valley, but that the valley lands could be irrigated by gravity canals. It was decided to build at Laguna, about ten miles northeast of Yuma, a dam of the Indian (or floating) weir type which had been used successfully in India and Egypt under similar conditions. The contract for the dam was awarded to the J. G. White Company on June 6, 1905, at a bid of $797,650.[66] Work was begun on July 19, 1905; but in August 1906 the contractors petitioned for a release, and received it on January 23, 1907. About 34 per cent of the work had been done by the contractors. The work was resumed by the Reclamation Service and completed on March 20, 1909. The dam raises the water level about ten feet. It is 4,800 feet long and 244 feet wide. The total cost of the dam was $1,672,168,

struction of expensive irrigation works was found to be, in most cases, an unsuccessful investment. Therefore the only solution to the problem was for the government itself to undertake the reclamation of its lands. Instead of providing a specific appropriation from the Treasury for this purpose, however, Congress decided that the money for the reclamation of the lands should be supplied by the government if the settlers who reaped the benefits would agree to repay the money. This would be, in effect, a loan without interest. The Reclamation Act (*United States Statutes at Large*, Vol. XXXII, p. 388) was approved on June 17, 1902. Provision was made for the examination, survey, and construction of irrigation works required to reclaim the public lands; and for this purpose there were appropriated the receipts from the sale and disposal of the public lands in the arid region, this fund to be known as the "reclamation fund."

For a more detailed history of the Reclamation Act and the irrigation movement, see House Document No. 79, 57th Congress, 2d Session, *First Annual Report of the Reclamation Service* (1902), pp. 33–75.

[64] *Third Annual Report of the Reclamation Service* (1904), p. 193.

[65] *United States Statutes at Large*, Vol. XXXIII, Part 1, p. 224.

[66] House Hearings before the Committee on Irrigation of Arid Lands, 67th Congress, 2d Session, *Protection and Development of Lower Colorado River* (1922), Appendix A, p. 46.

but including some river bank revetment and the diversion works the cost amounted to about $2,100,000.[67] A secondary canal heads at the Arizona end of the dam and provides water for lands north of the Gila. The primary canal heads at the California end of the dam, follows the general course of the river to the south, crosses the Colorado at Yuma by means of an inverted siphon, and divides into two canals which provide water for lands between Yuma and the Mexican border. The siphon was a difficult and expensive piece of work. It is 930 feet in length and 14 feet in diameter, and was completed on June 29, 1912, at a cost of $677,648.

The first irrigation in the Yuma project under the Reclamation Service was by ditch in the season of 1907. With the completion of the siphon in 1912 some 60,000 to 70,000 acres of land could be irrigated by gravity, and the irrigation plan also provided for a pumping plant to irrigate from 40,000 to 50,000 acres of mesa land. The Yuma project is one of the most important projects undertaken by the Reclamation Service. Besides the features discussed above, the project includes 316 miles of canals and laterals, 16 miles of drainage ditches, and an extensive system of levees.[68] By June 30, 1935, the construction cost of the Yuma project had reached $9,373,126.[69] Like other projects along the lower river, the complete safety and development of this project depend upon river control and regulation.

Palo Verde Valley.—The Palo Verde Valley is located on the west side of the Colorado River in Riverside and Imperial counties, California. The valley is about 25 miles long and 6 miles wide and contains about 80,000 acres of irrigable land. On the south the Palo Verde Valley overlaps the Cibola Valley, and on the north it joins the Parker Valley. The Palo Verde Mesa is located immediately to the west, and still farther to the west, over a small divide, lies the Chucawalla Valley.[70]

About 1856 Thomas H. Blythe came to the Palo Verde Val-

[67] All-American Canal Board, *The All-American Canal* (1919), p. 21.

[68] *Fifteenth Annual Report of the Reclamation Service* (1916), pp. 69–70.

[69] *Annual Report of the Secretary of the Interior for the Fiscal Year Ended June 30, 1935,* p. 80.

[70] Senate Document No. 142, 67th Congress, 2d Session, *op. cit.,* pp. 55–56.

ley and acquired 40,000 acres of land which came to be known as the Blythe Rancho. The first Blythe water filing was made at Black Point on July 17, 1877.[71] Cattle raising, rather than agriculture, was the chief activity of the rancho, although Blythe did build a gravity intake and a main canal and laterals to irrigate a considerable area. In 1904 the Blythe estate was sold to Hobson and Murphy, who organized the Palo Verde Land and Water Company, a corporation which succeeded to the property and water rights of the Blythe Rancho. The corporation repaired and enlarged the irrigation system and promoted the colonization of the valley. In spite of a disastrous flood in 1905, the development of the valley continued. In 1908 the properties and rights of the Palo Verde Land and Water Company were taken over by the Palo Verde Mutual Water Company which made extensive improvements in the canal system. Land was taken up rapidly, but the real development did not come until after the completion of a branch of the Santa Fe Railway into the valley in 1915.

Practically all of the land in the Palo Verde Valley is in private ownership. The Blythe Rancho was sold off in small tracts averaging about 60 acres. In 1918 the Palo Verde Joint Levee District was organized for the purpose of flood protection and river control. At first this organization was separate from the mutual water district, but in 1921 both companies were under the same management. In 1923 the Palo Verde Irrigation District Act, creating the Palo Verde Irrigation District, was passed by the California legislature.[72] This special act was approved on June 21, and the organization was ratified by the landowners on October 27, 1923. In December 1924 the legality of the organization was sustained by the Supreme Court. The district includes practically the entire valley. It took over the properties of the Palo Verde Mutual Water Company and assumed its bonds of $350,000. It assumed also the bonds of the levee and drainage districts amounting to $2,443,330.[73]

[71] Subsequent filings were made in 1878, 1883, 1904, 1908, and 1911. These filings are duplicate and enlarged filings rather than cumulative filings. The first filings are prior to any others of which record has been found. See Adams, *op. cit.*, p. 330.

[72] *Statutes of California*, Forty-fifth Legislature (1923), pp. 1067–1116.

[73] Adams, *op. cit.*, pp. 328–30.

The Palo Verde Valley is at an elevation of from 240 to 275 feet above sea level. It is not likely, therefore, that the valley will ever be inundated permanently by floods from the Colorado River, the danger that threatens Imperial Valley. Like the Imperial Valley, however, the Palo Verde community has been a growing one with rising land values; and again, like the Imperial Valley, floods are its greatest threat. Past floods have caused extensive damage to the valley, and have run up an expense burden so great that some settlers have been driven from their homes. The future existence and prosperity of this community depend upon flood control.

Other irrigable areas.—West of the Palo Verde Valley in Riverside County, California, are the Palo Verde Mesa and the Chucawalla Valley. The Mesa is from 320 to 450 feet above sea level, and has about 18,000 acres of irrigable land. The Chucawalla Valley is an inland basin or sink, about 30 miles long and as much as 12 miles wide, which has no surface drainage outlet. It is at an elevation of from 360 to 450 feet above sea level. It contains about 44,000 acres of irrigable land. Between 1908 and 1920 a number of plans were devised to develop these districts, but the construction of the necessary irrigation works was never undertaken and the lands still lie unreclaimed.[74]

North of the Palo Verde Valley is the Parker Valley, sometimes called the Colorado River Indian Reservation project. The greater part of the valley lies on the east side of the Colorado River in Yuma County, Arizona; but a small part is located on the west side of the river in Riverside County, California. The valley is about 37 miles long and 7 miles wide. It is estimated that there are 110,000 acres of irrigable land in the project but 95,000 acres are subject to overflow during periods of maximum flood.[75] As early as 1863 an investigation was made of the valleys of the lower Colorado to select the one offering the greatest advantage as an irrigation project. The Parker Valley was chosen and was embraced in the Colorado River Indian Reservation established by act of Congress on March 3, 1865.[76]

[74] Senate Document No. 142, 67th Congress, 2d Session, *op. cit.*, pp. 59–60.

[75] *Ibid.*, pp. 52–55.

[76] *United States Statutes at Large*, Vol. XIII, p. 559.

A canal was built, but it proved to be unsuccessful and was abandoned in 1876. Since that time several surveys have been made, but no important developments have been achieved. Until adequate flood control and water storage could be provided on the river, the progress of reclamation in the Parker Valley would undoubtedly be slow.

The Mohave Valley is located principally in Mohave County, Arizona; but a small portion of the valley lies in San Bernardino County, California, and a still smaller portion in Clark County, Nevada. It is about 25 miles long and 5 miles wide, and contains about 27,000 acres of irrigable land. Reclamation began in 1891 when two acres of land were cultivated for a school garden near Fort Mohave. During the next few years a canal was built across the Indian Reservation and a levee system was constructed. However, the canal heading filled with silt, and when the levees were breached by a flood in 1914 they were not repaired.[77] The Mohave Valley repeats the story that is practically the same for all projects on the lower Colorado River. River regulation and flood control are essential if present property improvements are to be protected and if agricultural development is to continue.

FLOOD CONTROL

As the Colorado River comes out of the hills and onto the delta, it is naturally a quiet stream during nine or ten months of the year. During the other two or three months, the time of the spring floods, it becomes a powerful, turbulent river, and is a menace to all of the development along its banks. The Gila River is the most important tributary of the lower Colorado, and although it may be dry most of the time it has on occasions carried floods of huge proportions to the lower basin country. These floods are extremely flashy and short-lived, but they may produce higher flood stages and greater velocities than the more deliberate and longer-lived floods of the Colorado. Fortunately the Gila floods occur in the winter time, usually between November 30 and March 1, and have never coincided with the high water in the Colorado. If they had coincided, the menace

[77] Senate Document No. 142, 67th Congress, 2d Session, *op. cit.*, p. 50.

to Yuma and to the Salton Basin would have been intensified and the levees would probably have been overwhelmed.[78]

Many thousands of years ago, the Colorado extended its delta across the Gulf of California and formed a dam separating the northern part of the Gulf from the ocean. This building process has continued, and today the dam is nearly 100 miles wide. The course of the river lies between the crests of the delta cone and turns to the left to flow to the Gulf. To the right lies the Salton Basin, the old sea bed of the upper Gulf, the lowest point of which is more than 300 feet below the river level. The average fall of the Colorado from Yuma to the Gulf of California is about 1.25 feet per mile, while the average fall from the Colorado River to the Salton Sink is about three feet per mile. It is only natural that the river should tend to abandon the sluggish grade to the Gulf and follow the much steeper grade to the Salton Basin. It has been proved definitely that the Colorado River has flowed a number of times into the Salton Basin during the past 10,000 years. The river would build up its bed a little higher each year until, at some high water, it could break through its northwest bank and pour itself down the steeper grade into the desert. In 30 or 40 years the basin would be filled and would become an inland sea, probably with an outlet to the Gulf to carry off the high waters at flood time. The steeper grade would be gone, however, and the river would return to its old business of delta building. It would turn once more to the Gulf, down the only grade left to it. Gradually the dam would be repaired, and during the following years the inland waters would evaporate, permitting the basin to revert to desert. The Colorado would start building up its bed again, to prepare once more for the time when it would shift its course back to the Salton Basin.

It is estimated that less than a thousand years ago, the Colorado made its last shift from the basin back to the Gulf. For many years the basin had been filled. Indians had built villages on its shores and had supported themselves by fishing and by cultivating the overflow lands. When the waters receded the

[78] James H. Gordon, "Problems of the Lower Colorado River," *Monthly Weather Review*, Vol. LII, No. 2 (February 1924), p. 95.

Indians were forced to abandon their homes, but traces of the old villages still remain. Then followed a period when the basin became a sun-baked desert, one of the most desolate spots in the United States. It was considered a menace by early travelers, and at one time there was a bill before Congress to dredge a channel through the natural dam to permit the waters of the Gulf to flow into the basin and to cover the desert.[79] At another time it was proposed to make an artificial break in the north bank of the Colorado and to force the river into the basin again. Finally, an artificial break was made—not to flood the basin, however, but to bring water in for irrigation purposes. Thousands of acres of rich land were brought into production; settlers arrived to make their homes there; and thriving communities were built. Yet always in the background was the threat of the Colorado, which, after a thousand years of preparation, was ready to shift its course of its own accord back into the basin—not to blot out a desolate valley as in former times, but to destroy a rich agricultural empire.

The Colorado held its course to the Gulf along the east side of the delta without marked changes in position of the channel for at least 55 years between 1850, the date of the settlement of Yuma, and 1905, when the river flowed into Imperial Valley. It is probable that the course remained virtually unchanged for 300 years, although it is known that from time to time some flood waters found their way from the Colorado River to the bottom of the basin. A tradition among the Indians of the region indicates that quite a volume of water existed in the basin within comparatively recent times. The earliest known existence of the Salton Sea within historic times is that shown by Rocque's map, dated 1762.[80] Father Pedro Font explored the region in 1776 and doubtless crossed the desert at least twice. His map shows an irregular opening which may have been meant to indicate an inland sea. Then for some 50 years there is a gap in our knowledge of the Colorado Delta; but collated reports give

[79] United States Department of Agriculture, Office of Experiment Stations, Bulletin No. 158, *Annual Report of Irrigation and Drainage Investigations* (1904), p. 176.

[80] Map of John Rocque in the British Museum. See D. T. MacDougal and collaborators, *The Salton Sea; A Study of the Geography, the Geology, the Floristics, and the Ecology of a Desert Basin* (1914), p. 15.

the presence of flood water in some volume in the sink in 1828, 1840, 1852, 1859, 1862, 1867, and 1891.[81] In 1891 so much water flowed over the western banks of the river that it found its way through the dense bordering growth of brush and weeds and reached the lowest part of the Salton Sink. A lake of 100,000 acres was formed; but the flow was not of sufficient duration or in sufficient volume to effect a permanent channel change. It is interesting to note, however, that at this same time the plans for irrigating Imperial Valley were under preliminary discussion. Scientists who understood the danger signal warned the prospective settlers of the impending menace, but their warnings were not heeded.

In 1903 and 1904 only waste water from irrigation reached the Salton Sink, but in 1905 the Colorado River broke through the lower Mexican heading of the Imperial Valley Canal and eventually discharged its full volume into the Salton Sink by way of two of its old overflow channels, the Alamo and the New River. The flood was the direct result of carelessness on the part of the settlers and engineers interested in the Imperial Valley project, and should never have occurred. The original intake of the Imperial Valley Canal at Hanlon's Crossing, California, had become so silted that the canal could not deliver sufficient water for the needed fall irrigation. A steeper grade, which a heading in Mexico would have provided, was needed. In addition, the establishment of Mexican water rights was desired, since the federal government of the United States was contesting the right of the Imperial Valley to divert water from the Colorado, still technically a navigable stream. Therefore a new cut was made in Mexican territory at the lower Mexican heading of the canal, called "Intake Number 3." The cut was made in October 1904 under the supervision of Mr. C. R. Rockwood, engineer for the California Development Company. High water was not expected until the spring of 1905, and few precautions were taken against an unexpected rise even though the cut was made above a depression in a bank composed of light alluvial soil. The intention was to close the cut in February; but three unseasonal rises caused by floods in the Gila River and followed by exceptionally

[81] La Rue, *op. cit.*, p. 141.

heavy summer floods in the Colorado delayed the work. The soft soil was eroded easily by the floods, and the artificial cut widened and deepened rapidly to accommodate the entire flow of the river. On June 1, 1905, the river was discharging a total of 61,500 feet per second, of which 5,360 second-feet, or 9 per cent, entered the canal. On June 30, with the same river volume, 13,960 second-feet, or 22 per cent of the total flow, went down the old Alamo channel. On July 30 the canal was taking 86 per cent of the river flow, or 15,020 out of 17,500 second-feet; and on October 25 the full river discharge, which was then 6,000 second-feet, was going through the break. All attempts in 1905 to close the breach failed. Late in November a diverting dam of brush, wire, piles, and gravel, designed to throw the river away from the entrance to the canal, had been nearly completed when a flood from the Gila raised the Colorado some 14 feet in 24 hours, and swept the dam away. With the hope of diverting some of the flow back to the Gulf, a short channel was dug from the Alamo to the Padrones River to turn the water into Volcano Lake, a depression 10 miles long, 6 miles wide, and about 13 feet deep on the rim of the Salton Basin. A dam was built across the New River outlet of the lake so that the lake would discharge to the south and east into the Gulf, through an old channel called Hardy's Colorado, and not to the northwest through the New River channel leading to the Salton Sink. However, the Padrones broke a new channel across country north of Volcano Lake to the New River, from where it continued down this channel to its lowest level at the bottom of the Salton Basin.[82]

In the meantime the Southern Pacific Railroad Company had drawn on its resources to help to close the break, but by December 1905 the closure had not been made. The Southern Pacific was vitally interested in the safety of the Imperial Valley. Its main line tracks ran through the valley and were threatened with submergence by the rising waters of the Salton Sea. The railroad advanced money to the California Development Company to fight the river and received as security a majority of the stock of the company. In this way the Southern Pacific gained control of the California Development Company and took

[82] *Engineering News*, Vol. LV, No. 2 (February 22, 1906), p. 216.

charge of the situation. Rockwood resigned, and Mr. H. T. Cory took over the work in the spring of 1906 under a mandate from the railroad to close the break at any cost. Nothing could be done until the high waters had subsided, and the spring months were spent in gathering materials while the valley was forced to face another inundation. A part of Mexicali was undermined and destroyed, and Calexico was seriously threatened. The growing channels of the Alamo and of the New River destroyed many acres of rich agricultural land, and the Salton Sea drowned out the plant of the New Liverpool Salt Company which had been located on its shores. Cory's plan was to build a new head gate across the break. It was a practical plan, and his work was done carefully. Unfortunately a weakness developed in the head gate, and on October 11, 1906, two-thirds of the structure suddenly was swept away. It took three weeks more to close the breach, and not until November 4 had the Colorado been forced back to its old channel leading to the Gulf. Cory's victory over the river was short-lived. Levees were built to strengthen the closure, but on December 7 a flash flood from the Gila broke through the unfinished levees a half mile south of the cut. In less than two days the whole river was flowing once more into the Imperial Valley.

The valley was stunned by this unexpected turn in events. The situation was especially alarming, since the water flowing into the valley was cutting deep channels through the soft earth which were progressing rapidly upstream (back-cutting) in a series of cataracts. If these deep channels ever reached the Colorado the Imperial Canal intake would be at such a low level that it would be impossible to take the water by gravity to the basin, and the Imperial Valley, if not submerged, would revert to desert. An even more serious consequence, once the deep channels had reached the river, would be the extreme difficulty of turning the river back to its old course until after the entire Salton Basin had been filled. If the river could not be turned back, the channel of the main stream would be deepened ultimately to and beyond the town of Yuma. Such a development would destroy many homes and farms in that vicinity, would undermine the railroad bridge, and would imperil the government works at Laguna Dam. Besides, the Salton Basin would be submerged for many years, since the flood waters in a valley below sea level could not drain away and could escape only by evaporation. A conservative value of the property in the United States which was menaced by such a catastrophe is $200,000,000.[83]

The Southern Pacific Company was unwilling to stand the expense of another attempt to tame the river, and the federal government was asked to lend its financial aid. President Theodore Roosevelt, in order to avoid the delay of asking Congress for an appropriation, requested Mr. E. H. Harriman, of the Southern Pacific Company, to make the closure at government expense. At the time the Southern Pacific appeared to be the only agency equal to the task of controlling the river.[84] Harriman practically closed the Los Angeles and Tucson divisions of the railroad and brought piling, timbers, and other materials by special trains to the scene of the flood from as far away as New Orleans. The breach was finally filled on February 11, 1907, and this time the closure did not fail. Over $2,000,000 had been spent in closing these breaks and restoring the river to its old channel, and property damage of some $5,000,000 had been incurred.[85] Yet the threat of the Colorado was still present. A part of two seasons' overflow of the river had been sufficient to create an inland sea 76 feet deep and about 285,000 acres in area. The seriousness of the flood menace could be ignored no longer.

[83] House Document No. 359, 71st Congress, 2d Session, *Report of the American Section of the International Water Commission, United States and Mexico* (1930), p. 21.

[84] Senate Document No. 212, 59th Congress, 2d Session, *Imperial Valley or Salton Sink Region* (1907), p. 5. It is interesting to note that the Southern Pacific Company was not repaid by the government for the expense incurred in closing the break until 1930. The railroad presented its claim many times but was refused payment of the full amount on the ground that a large part of the property saved belonged to the company. Finally, on June 10, 1929, the Court of Claims of the United States rendered a judgment in favor of the Southern Pacific Company for $1,012,665. The appropriation was made and the judgment was paid on a War Department voucher on March 31, 1930. See House Document No. 246, 71st Congress, 2d Session, *Judgments Rendered by Court of Claims* (1930), p. 4. See also *United States Statutes at Large*, Vol. XLVI, Part 1, p. 125.

[85] Senate Document No. 867, 62d Congress, 2d Session, *Colorado River* (1912), p. 2.

The Imperial Valley was first protected from flood waters by the head-gate works of the canal and by an embankment on the river side of the canal. This levee became a secondary line of defense. After the break in 1905, in spite of the obstacles encountered owing to the international situation and the Mexican government's unwillingess to co-operate, men and materials were sent into Mexico to supplement it by stronger embankments. These levees were all built of the soft earth of the delta, the only material available. They were easily breached, even when faced with rock, and were not an adequate protection against a river as strong as the Colorado during its flood periods. They are all located entirely in Mexico, and protect lands both in Mexico and in the United States.

For three years, from 1906 to 1909, the Colorado built up its bed, and then left its old channel again at a point about 29 miles below Yuma. It turned west into another of its old overflow channels, called the Bee (or Abejas) River, and then on into Volcano Lake. The river deposited tremendous quantities of silt in the lake and raised the bed so rapidly that it became extremely difficult to keep the levee above it. The Colorado failed to reach the Imperial Valley in 1909, but it had made a major step closer to its goal. The valley's defenses were seriously weakened, and the federal government realized that prompt action must be taken. Congress appropriated $1,000,000 in June 1910, to be spent in turning the river back into its former channel and keeping it there.[86] In July, President Taft appointed Mr. J. A. Ockerson, member of the Mississippi River Commission and an expert on work relating to alluvial streams, to make an examination and to devise a plan for restoring the river to its former bed. The plan followed was to divert the river back to its old channel and to build a levee some 25 miles long near the west bank of the river to prevent any future breaks in that direction. As with other levees, the projected work was located in Mexican territory, and it became necessary to secure the consent of the Mexican government before active operation could begin. The diplomatic agreement was not consummated until early in January 1911, and by that time the low-water

[86] *United States Statutes at Large*, Vol. XXXVI, Part 1, pp. 883–84.

season had passed. To add to the complications, this section of Mexican territory was in the hands of revolutionists, who attacked workers, stole supplies, and interfered with the transportation of materials. The Ockerson Levee was completed in 1911, but the summer floods of that same year breached the levee at several points, the largest of which was at the entrance to the Bee River channel. The water widened this break rapidly, and soon the entire river was flowing once more down the Bee channel to Volcano Lake. Parts of the levee still remained intact, but the failure was so complete that nothing further was done to restore it. All hopes of confining the Colorado to its old channel were given up. An elaborate system of levees was built from Hanlon's Heading all the way across to Volcano Lake and on to the rising ground of the Cocopah Mountains. Every year these levees had to be heightened as the River deposited silt and raised its bed faster than dirt could be piled on top of the levees. In 1914 the Volcano Lake Levee was breached and 10,000 cubic feet of water per second flowed through the levee into the Imperial Valley for many days before the levee could be repaired. In 1915 Congress appropriated $100,000 to protect the Imperial Valley, this money to be spent under the direction of the Secretary of the Interior when the Imperial Irrigation District had raised a like amount. On March 8 the money was raised, and General W. L. Marshall, consulting engineer to the Interior Department, was sent to Imperial Valley to supervise the work, which consisted of several miles of revetment along the Colorado River and the heightening of the Volcano Lake Levee.[87] In 1917 two breaks occurred about 28 miles southeast of Calexico, and in 1918 and 1919 the river moved northward again to the Volcano Lake Levee. The water-soaked earth was so soft that trains could not be run along the top of the levee to carry on maintenance work in the usual manner. In 1921 the flood waters rose so high that they came within a fraction of an inch of overflowing. It is said that a wind which drove the water back a little is all that saved the

[87] *Engineering News,* Vol. LXXIII, No. 6 (February 11, 1915), p. 285; No. 10 (March 11, 1915), p. 510. See also *United States Statutes at Large,* Vol. XXXVII, Part 1, p. 861.

day. It became obvious that there was nothing to do but to try to keep the river out of the Volcano Lake settling basin. To save its life, Imperial Valley had to do what the federal government had failed to accomplish in 1911.

The new plan was not to restore the Colorado to its old bed, east of the Ockerson Levee, but to divert it to the drainage channels of the Pescadero, a tributary of the Hardy. During the winter of 1922, a channel was dug about eight miles east of Volcano Lake through the delta cone between the Bee and the Pescadero, and a rock-fill diversion dam was built at the same time. Diversion was accomplished by dumping blocks of rock into the river faster than the water could carry them away, rather than by building elaborate diversion works. Thus the river was turned away from Volcano Lake through the Pescadero Cut to the Gulf; but it was well known that this solution to the problem was only a temporary one. In June 1922 six miles of levee were destroyed and had to be rebuilt. The flood waters of 1922 and 1923 were carried successfully; but instead of scouring out a deep channel, the river was spreading out and building up its bed. Levees which held the river in 1922 but which were discarded for new lines were overtopped in 1923.[88] The grade of fall flattened, and on June 12, 1926, the river broke through the Pescadero Levee six miles below the diversion dam and, with a flow of only 70,000 second-feet, spread northward as far as the edge of the Bee River channel. If the flow had been greater, the Colorado could easily have turned into the Bee River, thereby reverting to the conditions of 1921 with the Volcano Lake Levee serving as the only barrier between the river and Imperial Valley. In June 1922, when the Pescadero Cut was put in service, a flood of 112,000 second-feet followed this route without overflow; but with the continuous depositions of silt, the capacity of the channel had shrunk. On February 21, 1927, a flash flood came down the river and reached a maximum of 100,000 second-feet. The flood subsided quickly, but for a time the entire river was passing through the unrepaired 1926 break in the Pescadero Levee. In 1928 the flood peak of 100,000 second-feet passed Yuma without reaching dangerous heights on the Im-

[88] Gordon, *op. cit.*, p. 97.

perial Valley levee system, but on several occasions the river had broken through the first and second lines of levees and had reached the third and last line of defense. In the Colorado, floods of over 200,000 second-feet (approximately the flow over Niagara Falls), although unusual, are not unknown. It is well known that the levees could not hold out for long against such a force. The only thing that could save Imperial Valley was the construction of a large storage and flood-control dam upstream in the canyon country. The resources of the valley were not equal to such a tremendous project, and besides the Salton Basin communities were in no position to solve the economic and political problems involved. Without federal aid the Imperial Valley could do nothing but fight a losing battle against the river and await her fate.

The Imperial Valley is not the only district in the lower Colorado region that has suffered from Colorado River floods. All projects and settlements along the lower river are subject to floods of serious proportions; but the inundations are not permanent as in the case of the Salton Basin, which lies below sea level. The exceptional danger to the Imperial Valley has attracted attention away from the flood threat to the town of Yuma and the Yuma Valley, to the Coachella Valley, to the Cibola and Palo Verde valleys, to the Parker Valley, to the town of Needles and the Mohave Valley, and to other developments along the river which are subject to overflow during periods of high water.

The safety of the Yuma Valley is connected directly with the safety of the Imperial Valley in that a prolonged diversion of the river to the Salton Basin would cause back-cutting up the Colorado and imperil the government works at Laguna Dam and the whole Yuma project. In addition to this danger, the Yuma Valley is subject to floods both from the Colorado River and from its tributary, the Gila River. As early as 1893 a levee less than a mile long was completed at a cost of about $10,000 along the easterly boundary of Yuma to protect the town from the floods of the Gila.[89] Between 1905 and 1908

[89] House Document No. 2, 54th Congress, 1st Session, *Report of the Secretary of War*, Vol. II, Part 5 (1895), p. 3278.

another levee was built by the United States government extending south from Yuma along the left (or east) bank of the Colorado River to the Arizona-Mexican border. During the flood of 1909–10, this levee was breached by bank erosion at a point about 12 miles south of Yuma, and from 1909 to 1912 a total of $240,000 was spent on levees to protect the Yuma project. The levees extended from Laguna Dam south along both sides of the river for a distance of 45 miles.[90] On January 22, 1916, a flood of over 200,000 second-feet came down the Gila and the town of Yuma and the surrounding territory were severely damaged. The levees were breached, and in the town the water stood four feet deep in the streets, destroying many old buildings in the lower portions of the town. The Arizona side of the Yuma Valley was flooded and the town of Araz and the Indian Reservation were completely inundated. In addition to the damage from floods the expense of levee maintenance was incurred so regularly that it came to be regarded as a permanent expense chargeable against the lands protected.

The general development of the Palo Verde District began after 1907, and in 1908 the Palo Verde Mutual Water Company was formed to construct an irrigation system and a levee system for the benefit of the entire valley. The old Blythe levees were repaired from the intake to a point east of the town of Blythe, and the valley passed safely through the flood of 1909. In 1910, however, it was discovered that the bed of the river had been raised and that the levees would have to be heightened and strengthened. In 1912 flood waters flowed over the levees and a stream of water swept across some of the best land of the valley. The levees were raised from time to time, but the valley suffered from a repetition of disastrous levee breaks. The costs of levee maintenance increased, and taxes rose correspondingly. In 1916 an attempt was made to secure financial aid from the federal government; but it was not successful, although similar aid had been granted to the Imperial Valley in 1910 and in 1914. The bed of the river continued to rise, and in 1922 a very severe flood occurred. Thirty thousand acres were inundated,

[90] *Eighteenth Annual Report of the Reclamation Service* (1919), p. 393.

and the water stood several feet deep in the town of Ripley. Some people, including reputable engineers, held the federal government responsible for this disaster, because, according to one theory, the rising bed of the river at Palo Verde was a direct result of the building of Laguna Dam by the federal government for the Yuma project. Many of the farmers were unable to meet their financial obligations, and left the valley, thus increasing the tax burden on those who stayed. The tax rate rose to $19.50 an acre, while the population dwindled from 5,000 to 3,000 and the cultivated area from 40,000 to about 25,000 acres. In addition, the rising bed of the river had caused a corresponding rise in the water table of the valley, and a drainage problem had developed.[91] Palo Verde could not wait for the completion of a large storage and flood-control dam. Financial assistance was needed immediately if the community then existing was to be saved. The federal government took cognizance of this situation in 1929 when the Secretary of the Interior appointed a board to survey certain projects, including the Palo Verde Valley. The board recommended that the federal government take over the maintenance of the Palo Verde levees and relieve the farmers of the burden. This report prompted Congress to pass an act, approved April 19, 1930, directing that an official survey be made of the physical and economic problems of the Palo Verde and Cibola valleys.[92] The investigations were made by engineers of the United States Reclamation Service, and again it was recommended that federal assistance be given to this community, especially since all losses and expenses of protective work since 1909 might be traced to the government project at Laguna Dam. The levees needed to be raised at least three feet, and the settlers were no longer able financially to do the work. Still nothing was done, and in June 1932 another flood broke through the Palo Verde levees. Once more federal aid was solicited, and finally on July 1, 1932, an appropriation of $50,000 for the repair of the Palo Verde levees was approved.[93]

[91] Edward F. Williams, George W. Scott, and L. A. Hauser, *Protection of Palo Verde Valley, Calif.*, pp. 18–19.

[92] *United States Statutes at Large*, Vol. XLVI, Part 1, p. 222.

[93] *Ibid.*, Vol. XLVII, Part 1, p. 535.

The Parker Valley, to the north, was also subject to overflow. Although the river channel was more stable there, the valley's development depended upon an expensive system of levees, unless regulation of the flood waters were provided by storage reservoirs. In the Mohave Valley, the lands were inundated to some extent by floods of as low as 25,000 second-feet. Even with flood control by storage, some levee system would probably be necessary to reclaim these lands. In June 1912 the 20-mile levee north of Needles broke. The town was damaged, and about 200,000 acres in the upper Mohave Valley were flooded.[94] A development company had constructed about ten miles of levee averaging four feet in height, and in 1912 and 1913 the Indian Service constructed five miles of levee. Both of these levees failed in 1914 and have never been repaired.

It is difficult to estimate with exactness the amount of money that has been spent on the control of the lower Colorado River; but the greater part of the expense has been incurred below Laguna Dam. Up to 1927 it has been estimated conservatively that river control work below Laguna Dam cost about $12,500,000.[95] From 1919 to 1925 the average annual cost of levee maintenance for Imperial Valley was $200,000, while that for the Yuma project was $86,500. After 1925 the expenditures continued. Since the building and completion of the chief Boulder Canyon Project structures, however, it has not been necessary to maintain the levees up to their previous strength.[96] In 1926 the Imperial Irrigation District had about 78 miles of protective levees in Mexico, and it had 60 miles of railroad, trains of dump cars, and other expensive equipment to maintain these levees. The Yuma project had over 40 miles of levees in Arizona and California and about 39 miles of railroad.

Above Laguna Dam the greatest expenditures for river con-

[94] F. H. Brandenburg, "Flood in the Colorado," *Monthly Weather Review,* Vol. XL, No. 6 (June 1912), p. 918.

[95] William Kelly, "The Colorado River Problem," *Transactions of the American Society of Civil Engineers,* Vol. LXXXVIII (1925), Paper No. 1558, p. 314. See also House Document No. 359, 71st Congress, 2d Session, *Report of the American Section of the International Water Commission, United States and Mexico* (1930), p. 184.

[96] To bring these figures up to date Colorado River control expenditures incurred by the Mexican interests should be included, as well as expenditures incurred by the American interests.

trol have been made in the Palo Verde Valley. By 1925 the valley had 28½ miles of protective levees, and up to 1931 the Palo Verde Irrigation District had spent some $3,000,000 for levees and river control.[97] To this must be added the $50,000 appropriated by the federal government in 1932 to repair the Palo Verde levees. In the Mohave Valley a development company spent some $450,000 in building levees and canals, and in 1912–13 the Indian Service built a levee at a cost of nearly $25,000.[98] Thus the major expenditures for river control above Laguna Dam amount to some $3,500,000.

The expenditures for river control have been only a part of the costs incurred as a result of Colorado River floods. Farm lands have been inundated; crops have been lost; level lands have been gullied; irrigation systems have been ruined; farm buildings and equipment have been destroyed; cities and homes have been damaged; and lives have been lost. In the Imperial Valley alone, the physical damage caused by floods is estimated to have passed the $25,000,000 mark. The New River and the Alamo channels cut through 50,000 acres of farm land valued conservatively at $100 per acre. One hundred and fifty thousand acres more of good agricultural lands are still under the waters of the Salton Sea, along with the plant of the New Liverpool Salt Company, which was damaged to the extent of $476,746. Through evaporation, the sea has shrunk in size; but the time was reached when replenishments by waste water from irrigation offset the annual losses by evaporation. The Salton Sea is now a permanent body of water. Crop and property losses are estimated at $3,000,000, and loss of land previously reclaimed at $2,000,000. In addition, flood danger had caused interest rates to rise and land values to drop.[99] Other projects along the lower Colorado have had similar experiences and have suffered similar losses but on a smaller scale. The communities of the lower Colorado could not continue to meet

[97] Williams, Scott, and Hauser, *op. cit.*, p. 2.

[98] House Document No. 396, 63d Congress, 2d Session, *Combined Statement of the Receipts and Disbursements, Balances, etc. of the United States during the Fiscal Year Ended June 30, 1913*, p. 127.

[99] Senate Hearings before the Committee on Irrigation and Reclamation, 69th Congress, 1st Session, *Colorado River Basin* (1925), Part 1, p. 26.

such losses. Without flood control some of the most valuable projects in the lower basin would have to be abandoned.

POWER DEVELOPMENT

The history of power development on the lower Colorado River before the Boulder Canyon Project is very brief, since the use of the waters of the lower river to generate electric power is a comparatively recent development. Some power was genrated along the river for local use, such as the plant at Holtville, California, which supplied the Imperial Valley. But the great power possibilities of the Colorado River were barely touched until the development at Boulder Canyon was begun. The chief markets for such power are rather far removed from the river and, in consequence, power projects had to wait until long-distance transmission was perfected. Today power is playing a most important part in the development of the river, since it is through power that river regulation and flood control are being financed.

Because of the interstate and international character of the Colorado River, it was obvious that unregulated development of the stream would lead to a conflict of the various political interests involved. In addition there was also the problem of possible conflict between irrigation and power interests. To solve all of these problems, Congress passed an act, approved August 19, 1921, authorizing the states of Arizona, California, Colorado, Nevada, New Mexico, Utah, and Wyoming to enter into a compact not later than January 1, 1923, and providing for an equitable division among those states of the water supply of the Colorado River and its tributaries.[100] To prevent possible duplication of investigations, the Federal Power Commission decided to wait until the formation of the commission under the act of August 19 before it made its own investigations and issued any power development permits of long duration. By 1921, eleven applications had been filed with the Federal Power Commission for sites on the Colorado River and its chief tributaries. One of the most important of these was the application of Mr. James B. Girand of Phoenix, Arizona, whose permit,

[100] *United States Statutes at Large*, Vol. XLII, Part 1, p. 172.

granted by the Department of the Interior, had been recognized by the Federal Power Commission when a preliminary permit was issued for a period of one year for a power project on the Colorado River near the outlet of Diamond Creek in the state of Arizona.[101]

As time went on, the complicated character of the development of the Colorado River was realized more and more clearly. The Girand permit was extended, but no permanent grant was made.[102] The suspension of action on all other applications was continued. This policy was still in force when, by joint resolution of Congress on March 4, 1927, the commission was directed not to issue or approve any permits or licenses affecting the Colorado River or any of its tributaries until the Colorado River Compact had been approved, or until March 5, 1929, if the compact had not been approved by that date.[103] By 1927 the Boulder Canyon Project had become an important issue in Congress, and the resolution of March 4 was passed to prevent the establishment of any rights on the river that might interfere with the proposed dam and power plant at Boulder Canyon. On December 21, 1928, the Boulder Canyon Project Act was passed, and under section six of this act the embargo on all Colorado Basin permits was continued (excepting on those for proposed power projects on the Gila River) until the act became effective.[104] On March 1, 1929, a Congressional joint resolution was passed forbidding again the issuance of permits affecting the Colorado River and its tributaries (excepting the Gila) until March 5, 1930, unless the Boulder Canyon Project Act should be approved before that date.[105] The act was approved by executive proclamation on June 25, 1929.[106] The embargo on the issuance of permits was lifted automatically, and full authority was restored to the Federal Power Commission.

By this time the commission had on file some forty-five sepa-

[101] *First Annual Report of the Federal Power Commission* (1921), pp. 32, 118.

[102] *Third Annual Report of the Federal Power Commission* (1923), p. 61; *Fourth Annual Report* (1924), p. 91; *Fifth Annual Report* (1925), p. 112; *Sixth Annual Report* (1926), p. 72.

[103] *United States Statutes at Large*, Vol. XLIV, Part 2, p. 1456.

[104] *Ibid.*, Vol. XLV, Part 1, p. 1062.

[105] *Ibid.*, p. 1446.

[106] *Ibid.*, Vol. XLVI, Part 2, p. 3000.

rate applications which contemplated installations aggregating about 17,000,000 horsepower.[107] The commission was anxious to make some dispositions of this work, which had been delayed for so long; but various interests, particularly those in the upper-basin states, desired a further postponement until Arizona should sign the compact and until the upper-basin states should divide the upper-basin waters among themselves. It was proposed to make a study of the entire situation in an attempt to arrive at a settlement that would be satisfactory to the upper-basin states. As a result of this study, it was decided that the commission should withhold authorization for those projects which might affect interstate water allocations during the attempts being made by the respective states to negotiate agreements. Other applications, including small projects and those proposed on the Gila River, were to be disposed of after examination by the Department of the Interior as to their general effect on Colorado River development. Pending applications found not to be adapted to existing conditions were to be rejected. Under these classifications, the Girand application, after many postponements, was finally rejected.[108] The commission believed that this project would interfere with the interstate water allocation agreements, that Girand was not financially able to proceed with the project, and that Girand had not shown satisfactorily the justification of the project in the public interest.

Thus, with the rejection of one of the oldest applications for power development on the Colorado River, the Federal Power Commission indicated definitely that all power development on the river would be under its close supervision and that all power development would be carefully planned. Power had come to be of great importance to the entire Colorado River Basin. It is not only one of the greatest resources of the basin but is also the source of the wealth that is giving that region the river regulation and flood control that are so vital to its future development. Haphazard or unwarranted and wasteful power development could not be tolerated.

[107] *Tenth Annual Report of the Federal Power Commission* (1930), p. 9.

[108] *Thirteenth Annual Report of the Federal Power Commission* (1933), pp. 227–28.

Chapter II

THE COLORADO RIVER COMPACT

DRAFTING THE COMPACT

The Colorado River Basin comprises some 244,000 square miles and embraces parts of seven states and of northern Mexico. This huge area is divided into two sections by a 400-mile gorge, the northern end of which is at Lees Ferry, Arizona.[1] Above this point lies the upper basin of the river in the states of Colorado, New Mexico, Utah, and Wyoming. Downstream lies the lower basin in Arizona, Nevada, California, and northern Mexico. The two basins are about equal in area,[2] but there is far more irrigable land in the upper basin than in the lower.

Before 1918 water appropriations were permitted with very few restrictions. Development in the lower basin had proceeded at a much more rapid rate than in the upper basin, however, and the upper states feared that in view of the limited supply of water in that arid region the lower basin, through appropriation and use, would establish a legal title to such a large share of the water that the future development in the upper basin would be arrested. In addition it was contended that under the prevailing water laws of the arid West the construction of any large works on the lower river would serve to appropriate for use in the lower basin the entire available supply of water.[3]

[1] See Figure 1, facing p. 2.

[2] See Table III, p. 7.

[3] There are two distinct systems of water law in force in the United States. Where water is abundant, as in the Eastern states, the riparian system based upon the common law of England is used. Under this system ownership of the banks of the stream is the controlling factor. In the arid West, however, the doctrine of prior appropriation and use, derived from the laws of Spain and Mexico, is followed. Under this system, the first in use is the first in right, regardless of location on the stream. Of the seven states most interested in the Colorado River Compact, six (Arizona, Colorado, Nevada, New Mexico, Utah, and Wyoming) follow the doctrine of prior appropriation and use. The seventh state (California) recognizes a combination of the two systems. That is, she has sought to retain a modified common-law doctrine of riparian

The lower basin feared, on the other hand, that any rapid development of the great irrigable areas in the upper basin, which is the source of most of the water, would decrease the flow of water in the lower river so greatly that some lower-basin lands would be forced to revert to desert.

TABLE VI.—ACREAGE IRRIGATED IN COLORADO RIVER BASIN, BY POLITICAL BOUNDARIES*

	Irrigated 1920	Additional Possible	Total
United States			
Wyoming	367,000	543,000	910,000
Colorado	740,000	1,018,000	1,758,000
Utah	359,000	456,000	815,000
New Mexico	34,000	483,000	517,000
Arizona	501,000	676,000	1,177,000
Nevada	5,000	2,000	7,000
California	458,000	481,000	939,000
Total, United States	2,464,000	3,659,000	6,123,000
Mexico	190,000	610,000	800,000
Grand total	2,654,000	4,269,000	6,923,000
Total:			
Upper basin	1,530,000	2,550,000	4,080,000
Lower basin	1,130,000	1,720,000	2,850,000
(Gila basin)[a]	(430,000)	(400,000)	(830,000)
Total	2,660,000	4,270,000	6,930,000

* Senate Document No. 142, 67th Congress, 2d Session, *Problems of Imperial Valley and Vicinity* (1922), p. 33.

[a] Not added into total.

As a matter of fact, the upper-basin states had much more reason than the lower-basin states to be concerned about the legal status of water rights in the Colorado River region. The principle of prior appropriation and use had been administered by state courts in the Colorado River Basin within their respec-

rights and at the same time to enact a code of laws recognizing appropriations by beneficial use. See the respective state constitutions as follows: Arizona, Article XVII, Section 1; California, Article XIV, Section 3; Colorado, Article VI, Section 6; New Mexico, Article XVI, Sections 2, 3; Utah, Article XVII, Section 1; Wyoming, Article VIII, Section 3; and *Statutes of the State of Nevada* (1913), chap. 140, p. 192, Sections 2, 3.

tive jurisdictions for many years; but until 1922 the law was not at all clear as to the respective rights of the states to the waters of interstate streams. In 1907, in the case of *Kansas* v. *Colorado*,[4] the Supreme Court had indicated that a principle of "equitable division" would be followed rather than a principle of prior appropriation; but that case involved a controversy between one state which followed the common-law rule of riparian rights and another state which recognized the doctrine of prior appropriation. In 1922 the Supreme Court took a different position as far as water controversies between two prior appropriation states were concerned when it decided in the case of *Wyoming* v. *Colorado* that such water rights would be determined on the basis of prior appropriation and use regardless of state lines.[5] In other words, water users in California or Arizona were protected against developments in Colorado or Utah which would deprive the former of their supply. This decision confirmed the fears of the states of the upper basin. Naturally they would oppose the construction of large flood control and storage works on the river for the benefit of the lower basin, since that would speed the development of the lower states and, through regulation of flow, would permit them to appropriate more water. In addition, the use of the falling water to generate electric energy might be construed as a beneficial use, and then the upper basin would lose forever all rights to increased appropriations from the river. Without the support of the upper states, however, the lower states had little hope of securing through federal action the flood control and water storage so essential to the safety of their property and to their future development. Since the federal government appeared to be the only agency capable of coping with the many problems involved in the project, it was obvious that an agreement to secure an equitable division of the water would be advantageous to the lower basin as well as to the upper basin.

To reach such an agreement some informal discussions were held in San Diego, California, and in Tucson, Arizona, in 1918; and in January 1919 the Governor of Utah called a conference

[4] 206 U.S. 46.

[5] 259 U.S. 419.

of the seven Colorado River states to meet in Salt Lake City.[6] This meeting resolved itself into a permanent organization called the League of the Southwest. Resolutions were adopted to the effect that because of the magnitude of the problem and its interstate and international aspects, the federal government should lend its aid in developing the Colorado River Basin. Similar resolutions were adopted at a meeting held in April 1920 in Los Angeles. From August 25 to August 27, 1920, the League met in Denver, Colorado, to discuss the desirability of a large reservoir in the canyon country of the Colorado River for flood-control, irrigation, and power purposes. Arthur P. Davis, director of the United States Reclamation Service, attended the conference and assured the representatives of the seven states that there was sufficient water for both upper and lower basins and that the building of a reservoir on the lower river would not endanger the future of the upper basin. With this understanding in mind, the League adopted further resolutions demanding rapid development of the Colorado River Basin and asking that the waters of the river system be divided by compact. The seven states were requested to appoint commissioners for the purpose of drawing up the agreement for ratification by the legislatures of the states and by Congress.

Between February 22 and May 12, 1921, acts were passed by the seven states providing for the appointment of the commissioners.[7] On May 10, the governors of the seven states, or their representatives, met in Denver, Colorado, and formulated resolutions calling upon the President of the United States and Congress to grant the necessary consent to a compact between the states.[8] On August 19 the act of Congress to permit the com-

[6] House Hearing before the Committee on the Judiciary, 67th Congress, 1st Session, *Granting the Consent of Congress to Certain Compacts and Agreements between the States of Arizona, California, Colorado, Nevada, New Mexico, Utah, and Wyoming* (1921), pp. 22–23.

[7] Arizona, March 5, 1921, *Laws of Arizona* (1921), pp. 53–55; California, May 12, 1921, *Statutes of California* (1921), pp. 85–86; Colorado, April 2, 1921, *Laws of Colorado* (1921), pp. 811–15; Nevada, March 21, 1921, *Laws of Nevada* (1921), pp. 190–91; New Mexico, March 11, 1921, *Laws of New Mexico* (1921), pp. 217–22; Utah, March 14, 1921, *Laws of Utah* (1921), pp. 184–85; Wyoming, February 22, 1921, *Laws of Wyoming* (1921), pp. 166–67.

[8] The Constitution of the United States provides in Article I, Section 10, Paragraph 3, that: "No State shall, without consent of Congress, ... enter into any agreement or compact with another State, ..."

pact was approved.[9] This act provided that the compact should not become binding until approved by the legislatures of the seven states and by Congress. Provision was also made for the appointment by the President of the United States of a suitable person to meet with the commissioners of the seven states. The fact that the Colorado River is an interstate and international stream and that, at least by treaty with Mexico, it was still regarded as navigable, made it imperative that the federal government should be represented during the negotiations leading to a division of its waters. In addition, the various conflicting uses of the river, such as irrigation, power generation, and domestic water supply, indicated that federal supervision would be necessary. On December 17, 1921, President Harding appointed Mr. Herbert Hoover, then Secretary of Commerce, as federal representative on the Colorado River Commission.[10]

The first meeting of the commission was held in Washington, D.C., on January 26, 1922. Mr. Hoover was elected permanent chairman, and the work of drafting a compact was begun. Several meetings were held in Washington between January 26 and January 30, but no agreement was reached. On March 15 a meeting was held in Phoenix, Arizona, to give the state of Arizona a full opportunity to present her views, and subsequent meetings were held in El Centro and Los Angeles, California; Salt Lake City, Utah; Grand Junction and Denver, Colorado; and Cheyenne, Wyoming. On November 9, 1922, the final session of the commission was begun at Santa Fé, New Mexico. An agreement was reached on November 24, and the compact was signed by the representatives of the seven states and by Mr. Hoover as representative of the United States.[11]

ANALYSIS OF THE COMPACT

Considering the importance of the Colorado River to the Southwest and the many complications involved, the Colorado River Compact was a remarkably concise document and its

[9] *United States Statutes at Large*, Vol. XLII, Part 1, pp. 171–72.

[10] *Engineering News-Record*, Vol. LXXXVII, No. 26 (December 29, 1921), p. 1076.

[11] House Document No. 605, 67th Congress, 4th Session, *Colorado River Compact* (1923), pp. 1–2.

language was simple and clear.[12] The preamble set forth the desire of the seven states to compact, and cited the Congressional act of August 19, 1921, in which the states were given the permission to enter such an agreement. The state commissioners and Mr. Hoover, representative of the United States, were named as having agreed to the eleven articles of the compact.

Article I outlined the major purposes of the compact, which were seven in number. The first was to provide for the equitable division and apportionment of the use of the waters of the Colorado River system. Thus, if such an agreement were reached, the rapid development of the lower basin would no longer be a source of concern to the upper basin, since the lower basin could not establish a right to any more water than the amount apportioned to it under the compact. In other words, the compact would limit the operation of the prevailing water laws of the arid West, which provided for the establishment of water rights on the basis of prior appropriation and use. The second major purpose was to establish the relative importance of the different beneficial uses of the water. Later, under Article IV, *b*, the use of water for the generation of power was declared to be subservient to the uses of water for agricultural and domestic purposes. The third and fourth major purposes were to promote interstate comity and to remove causes of present and future controversies. The very fact that the compact existed would prevent disputes which might otherwise arise. In addition, Article VI provided for a commission to hear and adjust controversies between the states concerning water rights. The fifth, sixth, and seventh major purposes of the compact were to secure expeditious agricultural and industrial development of the Colorado River Basin, to secure the storage of waters, and to secure the protection of life and property from floods. The statement of these three factors as major purposes indicated definitely the point of view that the Colorado River Basin should be considered as a whole, and that one section was not to be permitted un-

[12] The original compact is filed with the Secretary of State of the United States and is not available to the writer. The copy of the compact used in this analysis may be found in Ray Lyman Wilbur and Northcutt Ely, *The Hoover Dam Power and Water Contracts and Related Data* (1933), pp. 387–90.

warranted development to the detriment of another. It was realized that to secure the most advantageous development of the entire basin, future needs and possibilities would have to be considered, and that each step would have to be taken in accordance with a carefully devised plan. However, the Colorado River drains a large area embracing widely different types of country. The problems of the northern portion are not the same as the problems of the southern part. To meet this situation, the two sections were given specific allocations of water so that each could work out its unique problems independently and without conflict with the other.

Article II defined the more important terms used in the compact. The term "Colorado River System" was defined as that portion of the Colorado River and its tributaries within the boundaries of the United States. Thus the compact was not concerned with merely the main stream bearing the name "Colorado River" but included within its scope all of the streams within the United States which form a part of the Colorado drainage basin. The term "Colorado River Basin" was next defined as the drainage area of the Colorado River system and all other territory within the United States to which the waters of the river could be beneficially applied. In later paragraphs of this Article, all lands in the upper- and lower-basin states outside of the true drainage basin which are served by diversions from the river were again specifically included as a part of the Colorado River Basin. The importance of this distinction is that districts such as the Imperial Valley were brought within the scope of the compact. By tunneling through mountains, it is possible that great additional areas of land may be covered by the compact; but it should be noted that any diversions from the Colorado River system to lands in states other than those specified would not be included as a part of the basin under the terms of the compact.

In paragraphs *c* and *d*, Colorado, New Mexico, Utah, and Wyoming were grouped together as the "States of the upper division," and Arizona, California, and Nevada were combined as the "States of the lower division." Lees Ferry was located as a point on the main stream one mile south of the mouth of the

Paria River, and the portion of the basin above this point was designated as the "Upper Basin," while the portion below was defined as the "Lower Basin." In defining the upper and lower basins of the Colorado in this manner, state lines were ignored, and small sections of Arizona were included in the upper basin while parts of Utah and New Mexico fell within the boundaries of the lower basin. For most purposes, however, the "States of the upper division" may be said to correspond with the "Upper Basin," and the "States of the lower division" to correspond with the "Lower Basin."

The final definition in Article II was of the term "domestic use." The uses of water for household, stock, mining, milling, industrial, and other like purposes were included; but the use of water for the generation of electric power was specifically excluded.

Article III stated definitely how the waters of the Colorado were to be divided between the two basins. The upper and lower basins were each given in perpetuity the exclusive beneficial consumptive use of 7,500,000 acre-feet of water per annum, which was to include any rights already established. In addition, the lower basin was given the right to increase its beneficial consumptive use by 1,000,000 acre-feet per annum if sufficient water existed in the stream to warrant the increased appropriation. Only water that is consumed was included in these quotas. The term "beneficial consumptive use" did not mean water that is diverted and later returned to the river. It meant the amount of water lost to the river during the uses of the water diverted, or the difference between the aggregate diverted and the aggregate return flow.

The states of the upper division were forbidden to permit the flow of the river at Lees Ferry to be depleted below an aggregate of 75,000,000 acre-feet for any period of ten consecutive years, thus fixing a definite quantity of water which must pass that point. Although the lower basin had been already guaranteed 7,500,000 acre-feet a year (with an increase to 8,500,000), this provision was adopted as a precaution to assure the lower basin first call on water up to a total of 75,000,000 acre-feet each ten years. It should be noted that water of the tributaries in the lower basin was not included in the 75,000,000 acre-feet. Thus the lower-basin tributary water remained available for use over and above that amount, even though it was previously included in the 8,500,000 acre-feet yearly grant. No apportionment among the states was made; hence decisions concerning definite allocations of water to certain states could not be reached until future compacts were agreed upon among the upper-basin states and among the lower-basin states.

From the figures stated it appears to have been assumed that the average flow of the river was at least 16,000,000 acre-feet per annum. The authors of the compact believed that even more water was available, but they wished their estimates to be conservative. As a matter of fact, it appears now that the figures of the Bureau of Reclamation were optimistic and that there is not as much water in the river as originally supposed. According to more recent estimates, however, there is still believed to be sufficient water to carry out the terms of the compact.[13]

Provision was made for a possible agreement between the United States and Mexico in which the United States would recognize Mexican rights to the use of water from the Colorado River system. According to the compact, water granted to Mexico would have to come from the surplus waters not already apportioned to the two basins in the United States. If there should still be a deficiency in the fulfillment of the international obligation, the burden of such deficiency would have to be borne equally by the upper and lower basins.

The concluding paragraphs of Article III forbade the states of the upper and lower divisions to withhold or require water which they could not reasonably use. A further equitable apportionment of the excess waters not apportioned by this compact could be made after October 1963, but only if one of the basins had reached the total beneficial consumptive use of the waters allotted to it. To make the further apportionment, a commission had to be appointed by the governors of the seven states and by the President of the United States to draw up a new agreement which would be, in effect, a new Colorado River Compact.

[13] Senate Document No. 186, 70th Congress, 2d Session, *Colorado River Development* (1929), pp. 44–51.

Article IV stated the preferred uses of the waters of the Colorado River, and the right of control by any state within its boundaries of the appropriation, use, and distribution of water was upheld. The waters were to be used for domestic, agricultural, and power purposes, and it was explicitly stated that since the river had ceased to be navigable for commerce, navigation should be subservient to the other uses. It should be noted also that power was declared subservient to agricultural and domestic uses. This condition was specifically included, because it is possible, although not probable, that at some lower dam at some particular season of the year it will be necessary to stop the entire flow of the river in order to create reserves for agricultural or domestic uses. This, of course, would stop temporarily the generation of power below that point and would result in a loss to the power interests. According to the terms of the compact, the power interests would have to accept such a loss without question. In the normal course of development of the river, however, it is to be expected that dams will be erected in both the upper and lower basins in such a way as to secure an average flow of water throughout the year and, with it, a maximum development of power. Nevertheless it must be admitted that the growing consumption of water in the upper basin will diminish the volume of power in the lower basin; and the upper basin cannot be denied its rights merely to satisfy the power demands of the lower basin.

Article V made the chief official of each of the seven states charged with the administration of water rights, the Director of the United States Reclamation Service, and the Director of the United States Geological Survey responsible for the systematic development of the Colorado River Basin. Statistics of stream flow at Lees Ferry were to be compiled, and other information was to be gathered and interchanged. Under this provision, information will probably be obtained which may point to a more equitable division of the Colorado River water if a further apportionment should ever be made.

Article VI provided the method for settling possible disputes between the signatory states. On the request of the governor of one of the states involved in the controversy, the states affected were to appoint commissioners with power to consider and adjust the dispute. The adjustment was to be subject to ratification by the legislatures of the states. It is only reasonable to expect that, as the resources of the Colorado River Basin are developed, points of disagreement between the policies of the various states will appear. Anticipating these disputes and providing for their settlement will undoubtedly save much controversy and prevent much ill feeling.

Article VII stated that nothing in the compact was to be construed as affecting the obligations of the United States to the Indian tribes. The reason for including this article is not clear, since there appears to be nothing in the compact that could possibly affect the rights of the Indian tribes. The importance of the article seems merely to be a definite restatement of the recognition of Indian rights.

Article VIII provided that present perfected rights to the beneficial use of waters of the Colorado River system were not to be impaired by the terms of the compact. Rights which were not perfected at the time of ratification of the compact were to be satisfied out of the water apportioned to the basin in which the rights were situated. The article also provided that whenever storage capacity of 5,000,000 acre-feet was provided on the main Colorado River within or for the benefit of the lower basin, present perfected rights in the lower basin would have first call on the water stored. Whether or not this storage takes place, however, the rights and obligations established under Article III were to remain in full force. It was believed that the entire low-water flow of the river had already been appropriated by users in California and Arizona, and these rights were upheld by Article VIII. If adequate storage were provided, the flow of the river would be regulated and disputes over low-water flow would cease. The 5,000,000 acre-feet of storage was considered ample for this purpose, but the statement of that figure could not be construed as a limitation on the maximum size of the reservoir to that amount. Actually, very large storage was necessary in order to protect the lower-basin states from the possible use by the upper-basin states of the entire flow of the river during periods of small runoff. With large storage capacity, the excess waters of the flood periods, which would otherwise be wasted into the sea, could be conserved and used to supply the lower basin during periods of low flow.

Article IX stated that nothing in the compact should be used to prevent a state from carrying on legal or equitable actions to protect rights under the compact. Thus the states can still decide disputes through the courts, although Article VI provided a method for the adjustment of differences through the appointment of commissioners.

Articles X and XI provided that the compact could be terminated at any time by the unanimous agreement of the signatory states and that it should become binding only when it was approved by the legislatures of the seven states and by Congress. Thus the unanimous consent of the interested parties was required before the compact could become effective. In addition it was definitely stated that in the event of termination all rights already established under the compact were to continue unimpaired.

THE ARIZONA CONTROVERSY

During the year 1923 the legislatures of California, Colorado, Nevada, New Mexico, Utah, and Wyoming ratified the compact.[14] Arizona alone failed to give her consent. Unfortunately the approval of the compact became a part of a political battle in Arizona, and the state ratification bill of 1923 was defeated by a narrow margin. In 1924 the Colorado River Compact was the chief issue in the Arizona political campaign. Governor Hunt, who had opposed the ratification of the compact, was re-elected, and the legislature took this as a mandate from the people to refuse to approve the compact under the prevailing conditions.

Arizona raised many objections to the compact and gave numerous reasons for her refusal to ratify the agreement. However, most of her fears were based upon imaginary difficulties or upon the conclusions hastily drawn from inaccurate or incom-

[14] *Statutes of California* (1923), chap. 17, pp. 1530–35; *Laws of Colorado* (1923), chap. 189, pp. 684–93; *Laws of Nevada* (1923), Resolution No. 2, pp. 393–99; *Laws of New Mexico* (1923), chap. 6, pp. 7–13; *Laws of Utah* (1923), chap. 5, pp. 4–10; *Laws of Wyoming* (1923), chap. 3, pp. 3–8.

plete information. Underlying all of the arguments that were brought forth in this controversy, Arizona seemed to be demanding two things: (1) a great quantity of water for her irrigable lands, and (2) revenue from hydroelectric power.

As noted above,[15] the Colorado River Compact did not divide the waters of the Colorado River system among the states. It apportioned a certain amount of water to the upper basin and a certain amount to the lower basin. By this arrangement sufficient water was reserved for the upper basin in spite of the rapid development of the lower basin, but as between the states of the lower basin the law of prior appropriation and use still held, Arizona feared that the rapid expansion of the use of water from the Colorado in California and in Mexico would establish rights to so much of the lower-basin allotment that there would not be enough water left to permit the future appropriations which would be necessary if her own development was to continue. Therefore she asked that the United States enter a treaty with Mexico limiting the amount of water that Mexico might have and demanded, before she would ratify the compact, that an agreement among Arizona, California, and Nevada be signed definitely dividing the lower-basin waters among the three lower-basin states. It is generally accepted that a treaty with Mexico to clear up the international situation in the lower Colorado River Basin would be desirable, and, as pointed out, an agreement among the states of the lower basin concerning a division of the water was contemplated by the compact. The difficulty was that the amount of water demanded by Arizona appeared to be unreasonably high; and no agreement could be reached. During the controversy the international situation faded from the picture and the real battle focused upon the division of water among the lower-basin states. Since Nevada had comparatively little irrigable land to which water from the Colorado River system could be advantageously applied, the two chief contestants were Arizona and California.[16]

Numerous proposals and counterproposals were made by Arizona and California in attempts to reconcile their differences;

[15] See discussion of Article III of the compact, pp. 62–63.

[16] See Table VI, p. 56.

but the two states seem never to have come close to an agreement. Arizona consistently demanded that only the main stream water should be considered, although this was contrary to the terms of the compact; and California consistently refused to yield on this point. According to one of the typical proposals[17] Arizona was willing to permit the allocation of 300,000 acre-feet per annum to Nevada, which was all Nevada would ever need, and 500,000 acre-feet to Mexico. This would leave 6,700,000 acre-feet of the 7,500,000 allocated to the lower basin, which Arizona contended should come entirely from the main stream. She proposed that the 6,700,000 acre-feet be divided equally between herself and California. Such an agreement would have left Arizona in sole possession of all of the water in the Arizona tributaries to the Colorado and would have added some 2,563,000 acre-feet to her quota. California did not consider this fair, since she wished recognition of 2,146,000 acre-feet of perfected rights and a diversion of 1,095,000 acre-feet for use by the Los Angeles Metropolitan Water District. When the 800,000 acre-feet allotted to Nevada and Mexico were added to these figures, the total came to 4,273,000 acre-feet, which left 3,227,000 acre-feet out of the total passing Lees Ferry. California proposed to divide this amount equally with Arizona and to permit Arizona to keep the water in the tributary streams. Such an arrangement would assign to California 4,854,500 acre-feet per annum as opposed to the 3,350,000 acre-feet she would secure under the Arizona proposition. All compromises suggested were refused by either one state or the other or by both.

In addition no agreement could be reached concerning the allocation of water to Indian lands. Arizona contended that the Indian reservations were entitled to a sufficient allowance of water to permit complete development of their irrigable lands, and she demanded that a share of the 7,500,000 acre-feet allotted to the lower basin should be reserved for use on the Indian lands.[18] California realized that such a plan would

[17] *Engineering News-Record,* Vol. XCVI, No. 1 (January 7, 1926), p. 36.

[18] Colorado River Commission of the State of California, *Colorado River and the Boulder Canyon Project* (1931), p. 285.

increase Arizona's share of the water, since the acreage of irrigable Indian lands is much greater in Arizona than in California. California refused to sanction the proposal, therefore, and took the stand that water for Indian lands should be included within the allotments made to the states.

The Arizona-California controversy continued for several years. Commissions met from time to time, but no agreement was reached. The general opinion seemed to be that Arizona's demands were exorbitant. But it should be remembered that the Colorado River is an important factor in the development of the state of Arizona. In the distant future Arizona must have access to great quantities of water if her mesa lands are to be cultivated. On the other hand, California had to have enough water for her perfected rights, for certain communities of southern California, and for the irrigation of land that can be cultivated more quickly and cheaply than the higher lands in Arizona. The Arizona proposals did not grant California enough water for her needs. It was known that the federal government would not build a large flood control dam in the canyon section of the river unless some sort of an agreement among the states could be reached. The flood danger to Imperial Valley and other California projects was becoming greater and greater as time went on, and the people of these communities were willing to accede to Arizona's demands, no matter how unreasonable, in order to get the dam built as quickly as possible. This proved to be a minority opinion in California, however, and in spite of the possible flood losses that might be incurred the state refused to accept the Arizona proposals.

The second underlying reason for Arizona's refusal to ratify the compact was her desire for revenue from the proposed power plant to be built, partly on Arizona's soil, by the federal government. Both Arizona and Nevada wished to be assured that their revenue from such a project would be approximately equivalent to the taxes which could be collected if the site, dam, and power plant were private property. Arizona argued that if coal, which might be used to generate power, could be taxed in Pennsylvania, and if petroleum could be taxed in California, Arizona should have the right to collect a tax on hydroelectric power, one of her greatest resources. Arizona wanted this right to be specifically recognized in any agreement between herself, Nevada, and California; but California refused to recognize such a right, since she would be the chief consumer of the power and would have to bear the burden of the tax. California felt that since the Boulder Canyon Project was to be financed chiefly through the sale of power, and since the main markets for the power were in California, she was bearing enough of the cost and should not assume a greater burden merely to give Arizona a revenue. California argued that Arizona would receive great benefits from the project which would be paid for by California's purchases, and that no more should be required.

Once again the points of view of the two states were in conflict. Eventually, however, California agreed to an allocation of power revenues both to Arizona and to Nevada. Arizona did not secure as large a revenue as she had demanded, but she did gain a right to certain revenues which are likely to represent a significant income to the state.[19]

THE SIX-STATE COMPACT

By 1925, it became apparent that Arizona and California were not going to reconcile their differences in the near future. Negotiations had been under way for two years, but agreement seemed still as remote as it had been in 1923. As hope for the ratification of the compact by all seven states faded, a six-state ratification movement was begun. It was proposed to waive the requirements of Article XI of the compact, which imposed the condition that the compact should be effective only when all seven states had ratified it, and to make the compact operative when six of the signatory states had waived the seven-state requirement, subject, of course, to approval from the Congress of the United States. During 1925, Colorado, Nevada, New Mexico, Utah, and Wyoming enacted legislation ratifying the six-state compact.[20] California also accepted the six-state

[19] See pp. 92–93, below.

[20] *Laws of Colorado* (1925), chap. 177, pp. 525–26; *Statutes of Nevada* (1925), Bill No. 87, pp. 134–35; *Laws of New Mexico* (1925), chap. 78, pp. 116–17; *Laws of Utah* (1925), chap. 64, p. 127; *Laws of Wyoming* (1925), chap. 82, pp. 85–86.

modification of the compact in 1925, but she accepted it with reservations. This ratification with reservations came to be known as the Finney Resolution.[21]

California feared that by accepting the compact she would lose her present perfected rights to water, which she had acquired through prior appropriation and use, in spite of the provision in the compact for the protection of such rights.[22] In order to be sure that an adequate water supply would be available to prevent impairment to her rights California demanded that immense storage be provided for the lower basin, and she felt that the 5,000,000 acre-feet of storage provided for in the compact was not nearly enough. In addition, California believed that under the six-state modification of the compact she would, in effect, be guaranteeing the performance of the compact agreement by the lower basin, since Nevada had comparatively little use for water from the Colorado River and Arizona could not be bound unless she tendered her ratification. California felt that she could not underwrite the compact for the upper basin unless she were confident she could meet all obligations; and, to be confident, huge storage facilities were necessary. Therefore, when California ratified the six-state compact in 1925, she attached the reservation that her ratification should not be effective until Congress authorized the construction of a dam of at least 20,000,000 acre-feet storage capacity. Thus from California's point of view, the compact would not become operative until further legislation had been enacted by the federal government.

The state of Utah had ratified the original compact in 1923 and the six-state condition in 1925; but early in 1927 its ratification of 1925 was repealed.[23] The reason given for this withdrawal from the six-state agreement was that Utah's rights to Colorado River water for irrigation and power development would be jeopardized, since Arizona, which had not ratified the compact, would be free to appropriate water and acquire priority over Utah. Later, however, the news spread that oil had been

[21] *Statutes of California* (1925), chap. 33, pp. 1321–22.

[22] See discussion of Article VIII of the compact, pp. 65–66, above.

[23] *Laws of Utah* (1927), chap. 1, p. 1.

discovered beneath the bed of the Colorado River in Utah, and Utah promptly passed a third act declaring the Colorado and Green rivers in Utah to be navigable streams and asserting the title of the state to the beds of those streams and all other navigable streams in the state.[24] Undoubtedly Utah expected to derive a considerable income from this oil development and wanted it clearly understood that if the Colorado River Compact should become operative it was not in any way to affect Utah's rights to revenue from her oil lands. A few months later, however, it was determined that the original reports concerning the oil discovery were false, and the oil field did not materialize. Therefore on March 6, 1929, Utah ratified the compact and accepted the six-state modification.[25]

In the meantime negotiations between Arizona and California had continued. At times the hoped-for reconciliation of differences was rumored to be very close, but on each occasion the rumors proved to be false. In 1925, after the failure of another meeting between the representatives of Arizona, California, and Nevada, the governors of the upper-basin states went on record as opposing the construction of any flood-control works on the Colorado until the compact was ratified and in force.[26] In California, in 1926, the legislature refused by an overwhelming majority to withdraw the reservations to the six-state ratification stipulated in the Finney Resolution.[27] Agreement on either a seven-state or a six-state basis seemed more remote than ever; but finally the federal government came to the conclusion that action should be postponed no longer. In 1928, in spite of Arizona's strenuous objections, Congress enacted legislation known as the Boulder Canyon Project Act, which was approved on December 21 of the same year.[28] It provided that the Colorado River Compact should become binding when it had been approved by all seven of the signatory states. If ratification could not be secured from all seven states within six months, however,

[24] *Laws of Utah* (1927), chap. 9, pp. 8–9.

[25] *Laws of Utah* (1929), House Bill No. 162, pp. 25–26.

[26] *Engineering News-Record,* Vol. XCV, No. 12 (September 17, 1925), p. 484.

[27] *Ibid.,* Vol. XCVII, No. 18 (October 26, 1926), p. 722.

[28] *United States Statutes at Large,* Vol. XLV, Part 1, pp. 1057–66.

the compact was to become binding when approved by six states, including California, and when these six states had waived the provisions of the first paragraph of Article XI which had required a seven-state ratification. The act provided also for a storage reservoir of a capacity of not less than 20,000,000 acre-feet. Thus the requirements of the California Finney Resolution, which had approved the six-state compact with reservations, were fulfilled, and the California ratification became binding. However, California enacted further legislation in 1929 ratifying the Colorado River Compact on the six-state basis.[29] Another provision of the Boulder Canyon Project Act provided that California must agree to limit her annual consumptive use of Colorado River water to 4,400,000 acre-feet plus one-half of any surplus waters unapportioned by the Colorado River Compact. California enacted the required legislation in 1929, and thus agreed to the water limitation.[30] Arizona still refused to ratify the compact, but with the approval of California, Colorado, Nevada, New Mexico, Utah, Wyoming, and the Congress of the United States, the Colorado River Compact became binding as provided for on the six-state basis.

For ten years Arizona remained outside of the interstate agreement, and all attempts to secure a ratification failed. In spite of this hostile attitude a provision was made in the Boulder Canyon Project Act for a water allocation to the state which was considered fair by all interested parties except Arizona. The act provided also for a revenue from the sale of power to Arizona and to Nevada. Arizona seemed to feel that the revenue should be greater; but in order for it to be greater, California, in return for flood protection, would be forced to contract for Boulder Canyon power at a price much higher than the estimated cost of steam power generated in California by the use of fuel oil. Such a move was considered to be poor business, since if California power interests should enter into such contracts (which was not likely) they would take only the minimum amount of Boulder Canyon power required and would generate

[29] *Statutes of California* (1929), chap. 15, pp. 37–38.

[30] *Ibid.*, chap. 16, pp. 38–39.

their own, cheaper power whenever possible. According to the general regulations for leases and contracts for Boulder Canyon power, 18 per cent of the total firm energy was reserved for Arizona and 18 per cent for Nevada for a period of 20 years.[31] So far, Arizona has not applied for any part of the power reserved for her, which indicates that Arizona has little use for the power at present and wanted control over it only to sell it at an advanced price to California. It appears, however, that Arizona's attitude toward the Boulder Canyon Project is changing. On March 3, 1939, Arizona ratified the Colorado River Compact on the condition that a tri-state agreement be concluded among California, Nevada, and Arizona for apportionment of the lower basin's share of the river's flow.[32] But California's Colorado River Board recommended to the state legislature the rejection of Arizona's proposal on the ground that Arizona had added new language to the tri-state compact, under the Boulder Canyon Project Act, greatly enlarging Arizona's rights.[33] Thus this latest effort to reconcile the differences between the two states appears to be doomed to failure, although Arizona's offer is a hopeful sign that some day the co-operation of all of the Colorado Basin states may be secured in the planning of projects designed to promote the development of the river.

[31] Ray Lyman Wilbur and Northcutt Ely, *op. cit.*, p. 105.

[32] *Engineering News-Record,* Vol. CXXII, No. 11 (March 16, 1939), p. 370.

[33] *Ibid.*, Vol. CXXIII, No. 3 (July 20, 1939), p. 61.

Chapter III

THE BOULDER CANYON PROJECT ACT

LEGISLATIVE HISTORY

During the explorations of the decade 1860–1870, when Major John Wesley Powell headed the first successful expeditions through the canyons of the Colorado, no thought seems to have been given to the possibility of a great reclamation project in the Colorado region. A few years later, however, Major Powell realized that the ultimate future of this section of the arid West depended upon the storage of water. He took up the problem of reservoir sites and convinced Congress that a survey should be made for the purpose of reserving the best locations. A survey was authorized, but unfortunately Congress decided after a short time that the project was not worth while and refused to continue the work. Nevertheless other people became interested in the Colorado River problem, and from the time of Powell on there was the gradual growth of a general plan for the regulation of the river, which soon included flood control and reclamation as important features. This became a definite program in 1928 with the enactment of the Boulder Canyon legislation. During the years that had passed it had become a far greater project than the pioneers had visioned and comprised one of the most comprehensive schemes for river development in the history of the world.

As early as 1907 the Boulder Canyon Project began to take form when President Theodore Roosevelt in his message to the Senate on January 12 described the disastrous flood in the Imperial Valley and outlined a rough plan of development under federal control.[1] Congress still believed that reclamation and flood control in the Imperial Valley should be left

[1] *Congressional Record*, 59th Congress, 2d Session, Vol. XLI, Part 2 (1907), pp. 1028–29.

largely to individual effort, and the President's recommendations were not followed. On February 16, 1918 (as previously discussed), the Secretary of the Interior and the Imperial Irrigation District entered into a contract to make a complete investigation, survey, and cost estimate of an all-American canal. On July 22, 1919, the All-American Canal Board, which had been appointed under the agreement of 1918, submitted a report, and recommended that an all-American canal be built and that large storage reservoirs be constructed by the federal government on the drainage basin of the Colorado River.[2] As a result of these recommendations, the first All-American Canal bill (also known as the first Kettner bill) was introduced in Congress on June 17, 1919.[3] Extensive hearings were held, but the bill did not come to a vote. On January 7, 1920, the second All-American Canal bill was introduced by Mr. Kettner, but it too failed to come before Congress for a vote.[4] Both bills had provided for the construction of an all-American canal, and the second had contained a provision for the construction of storage reservoirs large enough to provide an adequate supply of water for the reclamation of both private and public lands. Congress was not satisfied with the data available, however, particularly because no definite plan for storage had been devised. Therefore, on May 18, 1920, the Kinkaid Act was passed directing the Secretary of the Interior to make an examination of the Imperial Valley and to report on its condition and possible irrigation development.[5] The sum of $20,000 was provided for this investigation; but, as the work proceeded, donations from the Imperial Irrigation District, Coachella Valley, Palo Verde Valley, Arizona, Los Angeles, Pasadena, and other interested communities aggregating $171,000 were added to the fund, which, with further appropriations by Congress, finally reached a total of approximately $400,000.[6] A detailed investi-

[2] All-American Canal Board, *The All-American Canal* (1919), pp. 63–64.

[3] *Congressional Record,* 66th Congress, 1st Session, Vol. LVIII, Part 2 (1919), p. 1258.

[4] *Ibid.,* 2d Session, Vol. LIX, Part 2 (1920), p. 1204.

[5] *United States Statutes at Large,* Vol. XLI, Part 1, p. 600.

[6] House Report No. 918, 70th Congress, 1st Session, *Boulder Canyon Project* (1928), pp. 9–10.

gation was made, and the formal report was submitted to Congress on February 28, 1922.[7]

The recommendations of the report presented in definite form the chief aspects of the Boulder Canyon Project. Most of the recommendations were incorporated in the various Swing-Johnson bills, the last of which became the Boulder Canyon Project Act. The report recommended that a high-line canal be built by the United States from Laguna Dam to Imperial Valley, and that the cost be reimbursed by the lands benefited. Lands reclaimed were to be reserved for settlement by ex-service men under conditions securing actual settlement and cultivation. A storage reservoir was to be constructed at or near Boulder Canyon with government funds, and the government was to be reimbursed for this expenditure with the revenues derived from leasing the power privileges at the reservoir. The Secretary of the Interior was to be given the power to allocate to the various applicants their proportion of the power privileges and to allocate the costs and benefits of a high-line canal. The report suggested that all future developments on the Colorado River should be undertaken by the federal government and that in all construction and operation priority of right and use should be given first to river regulation and flood control, second to storage for irrigation, and third to power development.

To carry out the recommendations of the report similar bills were introduced at approximately the same time in the House of Representatives and in the Senate by Congressman Phil D. Swing and Senator Hiram W. Johnson.[8] These bills became known by the names of their co-authors, hence the name "Swing-Johnson." The first Swing-Johnson bill was introduced on April 25, 1922. It authorized the construction of the All-American Canal and of a dam at or near Boulder Canyon. It provided for the leasing of the power privileges by the Secretary of the Interior and stated that construction was not to begin until the lands to be irrigated were legally obligated to pay their proper pro-

[7] Senate Document No. 142, 67th Congress, 2d Session, *Problems of Imperial Valley and Vicinity* (1922).

[8] *Congressional Record*, 67th Congress, 2d Session, Vol. LXII, Part 6 (1922), pp. 5929, 5985.

portion of the cost. For the first time in the development of Boulder Canyon legislation, this bill authorized an appropriation ($70,000,000) to carry out the construction work.[9] Hearings were held on the bill, but it failed to come before Congress for a vote.

The second of the Swing-Johnson bills was introduced on December 10, 1923.[10] It was quite similar to the first Swing-Johnson bill but went into more detail regarding power leases. Like the first bill, the second failed to come to a vote. On December 21, 1925, the third Swing-Johnson bill was introduced, but on February 27, 1926, a new bill was substituted for it.[11] The third bill was very similar to the second except that it included a provision for a reservoir with a storage capacity of not less than 20,000,000 acre-feet. This provision not only satisfied the requirement of the California Finney Resolution, but also indicated definitely the tremendous size of the project. In addition the bill provided for a six-state ratification of the Colorado River Compact. The substitute bill increased the minimum size of the reservoir capacity to 26,000,000 acre-feet, and provided for an increase in the appropriation to $125,000,000, from which interest payments were to be deducted. The most important change, however, was the provision for the building of a unified power plant by the federal government in the place of the allocation of power privileges by the Secretary of the Interior as previously proposed.[12] It was believed that this arrangement would obviate controversies between applicants and long delays in their adjustment.

By the time the third Swing-Johnson bill and its substitute appeared in Congress for debate during 1926 and 1927 the Boulder Canyon Project had received country-wide attention, and had become an important national issue. However, the bill,

[9] The Colorado River Commission of the State of California, *Colorado River and the Boulder Canyon Project* (1931), pp. 70, 71, 74, 75, 76.

[10] *Congressional Record,* 68th Congress, 1st Session, Vol. LXV, Part 1 (1923), pp. 146, 217.

[11] *Ibid.,* 69th Congress, 1st Session, Vol. VI, Part 2 (1925), pp. 1232, 1323; Part 5 (1926), pp. 4683, 4730.

[12] The Colorado River Commission of the State of California, *op. cit.,* pp. 77–78, 80, 83, 84, 85.

like its predecessors, failed to come before Congress for a final vote; but this time the proposed legislation was defeated only after a bitter struggle in both the House of Representatives and in the Senate. The most spectacular part of the battle took place in the Senate, where, on February 22, 1927, the Boulder Canyon Project bill held the right of way over other legislation. The Senate was anxious to take up other business of importance, and a filibuster, led by Senator Ashurst of Arizonza, was used to displace the bill on the floor of the Senate. The Senate has a rule of unlimited debate, and to combat the filibuster the supporters of the bill managed to keep the Senate in session throughout the night of February 22, 1927. No adjournment was reached until the evening of February 23.[13] The filibuster was successful, but it was so sensational that it resulted in an avalanche of favorable publicity for the Boulder Canyon Project.

On December 5, 1927, the fourth and last of the Swing-Johnson bills was introduced by Mr. Swing in the House of Representatives, and on December 9 Mr. Johnson introduced a similar bill in the Senate.[14] The chief point of difference between this bill and the preceding Swing-Johnson bill was that the amount authorized for construction was increased to $165,000,000. At first it seemed as though the fourth Swing-Johnson bill also would meet with failure. The first session of the Seventieth Congress adjourned in May (1928) without taking action on the bill; but just before adjournment a joint resolution was passed providing for a thorough investigation of the economic and engineering features of the proposed dam by a board of five engineers and geologists.[15] The Secretary of the Interior appointed to the board Major General William L. Sibert, chairman, Charles P. Berkey and Warren J. Mead, geologists, and Daniel W. Mead and Robert Ridgway, engineers. This board ndered a report on December 3, 1928, which was very favorle to the project.[16] A few changes, which could be easily

[13] *Congressional Record,* 69th Congress, 2d Session, Vol. LXVIII, Part 4 (1927), 4396–4456; 4495–4563.

[14] *Ibid.,* 70th Congress, 1st Session, Vol. LXIX, Part 1 (1927), pp. 97, 341.

[15] *United States Statutes at Large,* Vol. XLV, Part 1, p. 1011.

[16] House Document No. 446, 70th Congress, 2d Session, *Report of the Colorado er Board on the Boulder Dam Project* (1928), pp. 1–15.

accomplished, were recommended; but aside from these qualifications the engineering and economic features of the project were endorsed unanimously by the board.

During the second session of the Seventieth Congress the Boulder Canyon bill was debated again in both houses. Once again Arizona fought the bill, but this time it came before Congress for a final vote and was passed by a large majority.[17] On December 19, 1928, the bill was sent to President Coolidge and on December 21 was approved by him. Arizona continued to withhold her consent; but by June 25, 1929, the requirements of a six-state ratification of the Colorado River Compact had been fulfilled. On that day the Boulder Canyon Project Act was made effective by President Hoover's proclamation.[18]

Thus after a long struggle, which may be traced back at least as far as the first All-American Canal bill of 1919, the Boulder Canyon Project received the sanction and support of the federal government. The local problem of the diversion of water from the Colorado River to the Imperial Valley over a route entirely within American territory had developed into a great project of interstate and international aspects, including water storage, flood control, river regulation, and power development. There were legal, political, engineering, and economic problems yet to be solved; but the first big battle had been won, and the project was now to become a reality.

ANALYSIS OF THE BOULDER CANYON PROJECT ACT

PURPOSES AND USES

The purposes of the Boulder Canyon Project as stated in the Boulder Canyon Project Act[19] were: (1) the control of floods; (2) the improvement of navigation; (3) the regulation of flow of the Colorado River; (4) the storage of water for the rec[lama]tion of public lands and other beneficial uses exclusively [within] the United States; and (5) the generation of electrical

[17] *Congressional Record*, 70th Congress, 2d Session, Vol. LXX, Pa[rt] pp. 603, 837, 897.

[18] *United States Statutes at Large*, Vol. XLVI, Part 2, p. 3000.

[19] *Ibid.*, Vol. XLV, Part 1, chap. 42, pp. 1057–66. Section 21 prov[ides] short title of the act shall be "Boulder Canyon Project Act."

These purposes seem to have been given approximately in the order of the priorities of their rights, since later in the act it was provided that as between the uses for the dam and reservoir of the project, preference is to be given, first, to river regulation, improvement of navigation, and flood control; second, to irrigation and domestic uses and the satisfaction of present perfected rights; and, third, to power. There is some possibility that the various purposes and uses of the project may come into conflict, and, therefore, it was necessary to establish preferences in order to avoid misunderstandings. For instance, room must always be left in the reservoir to take care of possible floods, although irrigation interests might desire that the reservoir be filled to capacity. Again, farmers of the lower basin may desire more water for irrigation at one season of the year than at another, although the generation of power can be maintained at a maximum only if a regular flow of water passes around the dam.

The preferences of the various uses given in the act do not correspond entirely with the preferences listed in the Colorado River Compact, although the Act is supposed to be subject to the terms of the compact. The reason for this discrepancy is that the act listed the uses for the dam and reservoir (or structures) of the project, while the compact dealt only with the uses for water. Thus the compact ignored flood control and gave first preference to agricultural and domestic uses. Flood control is not, in fact, a use of water as such; but, from the point of view of the project, it is the first and most important purpose of the legislation. If there had not been a desperate need for flood control, undoubtedly the Boulder Canyon Project would have been postponed indefinitely. A canal might have been built to the Imperial Valley entirely within American territory, but the construction of the huge dam and the reservoir in Black Canyon would have been long delayed.

The Colorado River Compact stated definitely that the Colorado River has ceased to be navigable to commerce and that navigation, therefore, should be subservient to other uses. Still the improvement of navigation was given in the act as the second purpose of the legislation, and it was listed as one of the preferred uses of the dam and reservoir. This inconsistency between the compact and the act is not likely to be important, since actual navigation is so small that a possible conflict between navigation and other uses is exceedingly remote. Probably the reason that improvement of navigation was given as one of the chief purposes of the act was to establish the authority of Congress to enact the legislation, for the federal government is empowered by the Constitution, under the application of the Commerce Clause, to carry on river work that is beneficial to interstate and foreign commerce.[20] In the case of the Colorado River the claim to authority to construct the project on the basis of improvement of navigation was questionable, since so much water will be diverted by the All-American Canal from the lower river that navigation below Laguna Dam will be practically impossible.

Regulation of flow was listed as the third purpose of the act, but it was the first mentioned in priority of uses. Regulation of flow was given this important position since it is through river control that the chief purposes of the act are to be achieved. By equalizing the flow, the high-flood peaks of the spring months will be eliminated, and the prime purposes of flood control will be accomplished. Lands previously subject to periodic overflow may now be brought safely under cultivation, and lands abandoned because of the flood danger may be resettled. Ordinarily regulation of flow would also be a boon to navigation, since the swift currents of the high-water periods and the shallowness of the stream during the low-water periods would be avoided. In this special case, however, as mentioned above, so much water is likely to be diverted from the stream that navigation below Laguna Dam will cease altogether, and regulation of flow could not possibly aid the situation. Actually the increased diversions from the river will be a direct result of river regulation, since the higher the minimum flow in the stream the larger the acreage that may be kept under permanent cultivation. Obviously such a condition is an aid to irrigation and reclamation, even though the navigability of the river is impaired. River regulation, of course, is an important factor also in the generation of electric power, in that the more nearly

[20] *Constitution of the United States*, Art. I, Sec. 8, Par. 3.

equalized the flow becomes the smaller the unutilized capacity of the power plant will be. Yet complete utilization of the power plant will never be achieved, since the stream will be regulated first for flood control, agricultural, and domestic purposes as indicated by the stated priorities.

The fourth purpose was the storage of water for the reclamation of public lands and for other beneficial uses exclusively within the United States. This purpose ranks second in the classification of priorities, since the term "other beneficial uses" includes water for domestic purposes[21] and for the satisfaction of present perfected rights as well as water for irrigation. The dam, reservoir, and canal will make possible great increases in the acreage of land cultivated in the lower basin of the river. The dam and reservoir will regulate the flow so that lands which could receive no water during the low periods of previous years may now be abundantly supplied and fully cultivated. The heavy silt load of the stream will be greatly decreased, and this troublesome and expensive element in the reclamation of lower-basin lands will be largely eliminated. The canal will bring water by gravity to Imperial and Coachella valley lands which formerly lay unused because of the actual lack of water or because the pump lifts from the old canal were too high. These benefits were meant to be enjoyed primarily within the United States, however. Benefits which may accrue to Mexican lands are incidental to, rather than a part of, the objectives achieved by water storage and river regulation. The act neither affirmed nor denied any Mexican rights to the waters of the Colorado, but the implication was that the application of the benefits of the project to Mexican lands was to be subservient to their application in American territory.

The generation of electrical energy was given the last place in the list of purposes of the act and last place in priority of uses. Yet in spite of this ranking it is one of the most important phases of the project. It is through the sale of electric power that the project is to be made a financially solvent and self-

[21] *United States Statutes at Large*, Vol. XLV, Part 1, Sec. 12, p. 1064, states that the term "domestic" used in the act shall include water uses defined as "domestic" in the Colorado River Compact.

supporting undertaking. The lands receiving the direct benefits of flood control, silt elimination, and irrigation could not bear the cost of the entire project; and the government was not willing to donate the money from the federal treasury without reimbursement. The only solution to this dilemma was to find a third method of financing the project, and the sale of electric energy which would be generated at the dam became the natural and logical answer to the problem. The Boulder Canyon Project has been accused by its opponents of being primarily a power project rather than a flood-control and reclamation project with power generation as an incidental, although important, feature. Whether this criticism is warranted or not will be discussed at greater length in a later section. The fact remains that no other practical method of financing the project had been suggested, and if the project could not have been made self-supporting through the sale of electric power it would not exist today.

HOOVER (BOULDER) DAM[22]

In order to achieve the purposes of the act, the Secretary of the Interior was authorized to construct certain works and make certain arrangements, including the construction, operation, and maintenance of a dam and incidental works in the main stream of the Colorado River at Black Canyon or Boulder Canyon adequate to create a storage reservoir of a capacity of not less than 20,000,000 acre-feet. Whether the dam should be built at Boulder Canyon or at Black Canyon made little difference to the

[22] During the past few years there has been an unfortunate controversy concerning the name of the chief structure of the Boulder Canyon Project, the great dam built in Black Canyon. In September 1930 Secretary of the Interior Ray Lyman Wilbur announced that the dam to be built in Black Canyon was to be called "Hoover Dam," in honor of President Hoover; and subsequently he issued an official statement to this effect. In May 1933, however, shortly after Mr. Roosevelt became President, Mr. Harold L. Ickes, who succeeded Dr. Wilbur as Secretary of the Interior, changed the name to "Boulder Dam," and since that time the structure has been referred to in the various acts of Congress and in the indexes to the Statutes of the United States as "Boulder Dam" or as "Boulder Canyon Dam." However, the writer can find no evidence that the change to "Boulder" was formally and officially made; and it appears, therefore, that the legal and proper name of the dam is still "Hoover Dam." For this reason the title "Hoover Dam" is used in this study, although the term "Boulder" has been inserted occasionally in parentheses to give recognition to the wide usage of that name for the structure. (See Homer Stille Cummings, *Selected Papers, 1933–1939*, pp. 254–55.)

project from the point of view of location of the dam site, since the two canyons are so close together that a dam in either one of them would serve the same purposes. The definite mention of the location of the dam in Boulder Canyon or Black Canyon, however, eliminated the consideration of other dam sites farther upstream or downstream, which, if utilized, might involve important changes in the plan of the project as outlined in the act and thus result in at least partial failure as far as the original purposes of the act were concerned. The majority of those who studied the various possible dam sites agreed that the best location for the dam under consideration from both the economic and engineering points of view was at Boulder Canyon or at Black Canyon, and the act was framed to develop the project according to this recommendation.

The California Finney Resolution was undoubtedly responsible for the definite limitation of the minimum size of the reservoir to 20,000,000 acre-feet. To create a reservoir of this size, a dam of a minimum height of some 500 feet would have to be built. With such limitations on minimum size the project became one of huge proportions, which involved engineering problems of unprecedented magnitude. According to the best engineering authorities, however, it was entirely feasible to construct a safe, practical dam which would fulfill the requirements of the project. As a matter of fact, the dam actually built is much higher and the reservoir created much larger than the minimum-sized dam and reservoir required by the act.

The building of such a huge dam and the storage of such a tremendous quantity of water naturally contemplated the reclamation of great tracts of land. The act provided that all public lands subject to practical irrigation and reclamation by the construction of the project were to be withdrawn from public entry; but this withdrawal was not to be permanent, since the Secretary of the Interior was given the authority to open these lands to entry in tracts not exceeding 160 acres in accordance with the provisions of the Reclamation Law.[23] These lands would have

[23] *United States Statutes at Large*, Vol. XLV, Part 1, Sec. 12, pp. 1063–1064, defines "Reclamation Law" as the act of June 17, 1902, and acts amendatory thereof and supplemental thereto. See *ibid.*, Vol. XXXII, Part 1, Sec. 4, pp. 388–90.

to bear their share of the cost of the irrigation works in relation to the benefits received, and the Secretary of the Interior was given the authority to allocate the burden and to specify the installments to be paid, again in accordance with the Reclamation Law. All revenues derived from this source were to be deposited in the Colorado River Dam Fund provided for in the Boulder Canyon Project Act. Thus the act made no provisions for land entry which come into conflict with the established precedents of the Reclamation Law. Actually the act leans heavily on the Reclamation Law for rules of land entry, and follows its provisions with but one exception. The lands made subject to reclamation by the Boulder Canyon Project were reserved to entry first by war veterans, and in addition preference was given to war veterans in all construction work on the project. This provision does not appear in the Reclamation Law, but it is supplementary to the law rather than in conflict with it and is applicable only to this one project. The preference was limited to a period of three months, and it was subject to the act of December 5, 1924, which provided that the ex-service men given preference must be sufficiently qualified for the work to give reasonable assurance of success.[24] The Boulder Canyon Project Act stated further that actual residence on the land by entrymen for one year is required and that lands relinquished before the end of the required period of residence will be reopened to entry at the end of sixty days. By these provisions the act was attempting to secure actual settlement of the lands and to prevent individuals from gaining control over large tracts of reclaimable land. Apparently the federal government intends to follow a more severe land policy in connection with the Boulder Canyon Project than it has in the past.

The act also provided that the title to the dam, reservoir, power plant, and incidental works of the project are to remain forever in the United States. Thus the Secretary of the Interior has no authority to transfer title to the dam, reservoir, or power plant, although the act does not state definitely whether or not this rule applies to the movable equipment in the power plant.

[24] *United States Statutes at Large*, Vol. XLIII, Part 1, Subsec. c, p. 702.

Provision was made for the leasing of some units of the power plant; but the federal government will keep permanent control over the irrigation and flood-control aspects of the Boulder Canyon structures. With federal control it is reasonable to expect that the power of river regulation will be used to promote the public interest and that it will not be used to cater to individual or sectional desires, a condition which might exist under private ownership.

ALL-AMERICAN CANAL

In addition to the construction of the dam and reservoir, the Secretary of the Interior was authorized to construct a main canal and appurtenant structures located entirely within the United States connecting Laguna Dam, or some other suitable diversion dam which the Secretary of the Interior may construct if necessary, with the Imperial and Coachella valleys in California. This provision gave federal sanction to the All-American Canal, the original project from which the larger and more inclusive Boulder Canyon Project developed. The building of this canal will release the American farmers in Imperial Valley from dependence upon water in a canal lying mainly within Mexican territory. The canal will enter the valley at a higher level and will make possible the irrigation by gravity of lands in both the Imperial and Coachella valleys which previously could not be reached by the waters of the old canal. Thus the canal will make possible the fuller and more economic development of the Imperial and Coachella valleys. Expensive pump lifts will be avoided, and a steadier flow of water may be diverted from the permanent diversion dam than was possible under the old system when temporary diversion weirs were built across the river during periods of low flow.

The cost of the construction of the canal and appurtenant structures is to be repaid to the federal government in the manner prescribed by the Reclamation Law.[25] That is, the repayment is to be made by the lands benefited according to the

[25] *United States Statutes at Large*, Vol. XLV, Part 1, Sec. 1, p. 1057; Sec. 4*b*, p. 1059.

benefits derived, as determined by the Secretary of the Interior, who must, by contracts or otherwise, insure the payment of all expenses of construction, operation, and maintenance of the canal and appurtenant structures.[26] These contracts are not to interfere with the contract dated October 23, 1918, between the United States and the Imperial Irrigation District, providing for a connection with Laguna Dam; but the Secretary was authorized to enter into further contracts with the Imperial Irrigation District or other districts, persons, or agencies to complete the financial plan of the canal. The Boulder Canyon Project Act provided that this plan had to be definitely completed before construction could start. In addition the financing of the All-American Canal was to be kept clearly separated from the financing of the dam and other works of the project. The reason for this distinction was that the All-American Canal is an irrigation and reclamation project within the scope of the Reclamation Law.[27] Although the money advanced by the federal government for such projects must be repaid, it is not to be repaid with interest. Therefore the All-American Canal, under the provisions of the Reclamation Law, will escape the interest burden, while the financial plan for the rest of the Boulder Canyon Project must include an interest charge. None of the revenues derived from the sale of power generated at the dam or from the sale of water for potable purposes outside the Imperial and Coachella valleys are to be used to pay for the canal. These revenues were reserved for the rest of the project, for which a different plan of financing was devised. No charge is to be permitted for water for irrigation and potable purposes in the Imperial or Coachella valleys, since the lands of these valleys will bear the burden of the cost of the canal. However, the districts irrigated by the canal are permitted to utilize the power possibilities of the canal

[26] *United States Statutes at Large,* Vol. XXXII, Part 1, chap. 1093, Sec. 4, p. 389.

[27] *United States Statutes at Large,* Vol. XLV, Part 1, Sec. 14, p. 1065, provided that the entire act is to be deemed a supplement to the Reclamation Law excepting as otherwise provided. Thus the Reclamation Law may be a great influence in all matters of construction, operation, and management of the whole project; but the greatest influence of the law will be in connection with the All-American Canal, which is specifically designated as a development coming under the provisions of the Reclamation Law.

and to use the revenue derived from the sale of that power in order to help repay their obligation to the federal government. In order to avoid any future controversies the power possibilities of the canal are to be divided between the districts in accordance with their shares of the cost of the canal.

When the entire cost of the canal has been repaid to the government, the Secretary of the Interior may transfer the title to the canal and appurtenant structures to the districts or agencies using the canal, excepting the title to Laguna Dam and the main canal to Syphon Drop. The reason for the reservation is that Laguna Dam and the first part of the canal to Syphon Drop are of importance to the whole lower basin below Laguna, while the rest of the canal is more of a local project. The transfer of the title to the canal from Syphon Drop on would be of interest only to the communities under irrigation from the canal. If and when the transfer takes place, each district or agency interested in the canal will receive a share in proportion to the capital investments made in the canal. The implication of these provisions is that the federal government does not intend to maintain its control over the entire All-American Canal, although it can maintain this control if it wishes to do so. It is likely that the section of the canal from Syphon Drop to the valleys will pass some day from federal ownership, but the rest of the Boulder Canyon Project will always be government property as expressly provided by the act.

POWER DEVELOPMENT

The discharge of water from the reservoir around the huge dam provided for in the Boulder Canyon Project Act will naturally create a tremendous quantity of energy which may be transformed into electric power. In order to reap the benefits of this situation and to provide a method for repayment of the cost of the dam and appurtenant structures to the federal government through the sale of power, the act authorized the Secretary of the Interior to construct, equip, operate, and maintain at or near the dam a complete power plant and incidental works suitable for the fullest economic development of electric energy from water passing around the dam.[28] The act was rather general in its requirements for power development in that it did not hold the Secretary of the Interior to a rigid plan so far as the operation of the power plant was concerned. To indicate several methods of procedure rather than to prescribe one set of rules and regulations was a wise move, since the best plan of construction and operation of the power plant could be devised more accurately after the requirements of the rest of the project were definitely known. The Secretary of the Interior was given three alternatives concerning the construction of the power plant and the disposal of the power generated at the dam: First, the government could build the powerhouse, install the machinery, operate the plant, and sell the generated power at the switchboard. Second, the government could build the powerhouse and install the machinery but turn over the operation of one or more units of the plant to lessees who would pay rent for the use of the powerhouse and of the machinery. Third, the government could build the powerhouse but permit lessees to install and operate the machinery. In the last case, payments made by the lessees would be based on a rental for the powerhouse and on the use of falling water.

Although these alternatives gave the Secretary of the Interior considerable leeway in fulfilling the requirements of the act for power development, they still confined him to a general method of procedure. No matter which alternative was selected, the rates charged for power or to lessees had to be reasonable, but also they had to be high enough to insure a return, which, with other revenues, would be sufficient to carry out the financial plan of the act. These rates were subject to readjustment periodically as justified by competitive conditions. Fifteen years after the date of execution of contracts for the sale of power or with lessees, and every ten years thereafter, the rates could be either raised or lowered upon the justifiable demand of either party to the contracts. Every contract was required to contain a provision under which disputes or disagreements could be settled either by arbitration or by court proceedings.

[28] *United States Statutes at Large,* Vol. XLV, Part 1, Sec. 1, p. 1057.

In the development of the power possibilities of the project and in all proceedings concerning the contracts the Secretary of the Interior was to be the representative of the United States government. This provision gave the Secretary very broad authority; but any rules and regulations prescribed by him had to conform with the requirements of the Federal Water Power Act. Thus in matters concerning the maintenance and repair of works,[29] the maintenance of a system of accounting, the control of rates and services in the absence of state regulation or interstate agreement, the valuation for rate-making purposes, the transfer and extension of contracts, the expropriation of excessive profits, the recapture and/or emergency use by the United States of the property of lessees, the Secretary must make his rules conform with the regulations laid down by the Federal Power Commission and must change his rules to conform with any future regulations which the Federal Power Commission may devise.[30] On the other hand, the Boulder Canyon Project Act limited the authority of the Federal Power Commission in that it forbade the approval of any power permits on the Colorado River or its tributaries (excepting the Gila, which joins the Colorado far below Boulder and Black canyons) until the act had become effective. In this way no power projects would be started which might conflict with the Boulder Canyon Project or which might become unnecessary and useless upon the completion of the dam and the power plant. The act limited the authority of both the Secretary of the Interior and the Federal Power Commission; and, in effect, it reserved the first choice of extensive power development on the Colorado River to the federal government. Therefore, although the generation of power is not one of the preferred uses of the project, it is obvious from the provisions in the act that the federal government intends to develop the power features of the project as

[29] *United States Statutes at Large,* Vol. XLV, Part 1, Sec. 12, p. 1064, defined the term "maintenance" to include provision for keeping the works in good operating condition.

[30] *United States Statutes at Large,* Vol. XLV, Part 1, Sec. 12, p. 1064, defined "Federal Water Power Act" as the act of June 10, 1920, and the acts amendatory thereof and supplemental thereto. See *United States Statutes at Large,* Vol. XLI, Part 1, chap. 285, pp. 1063–77.

fully and as effectively as engineering and economic conditions will allow.

FINANCIAL PLAN FOR THE DAM AND APPURTENANT STRUCTURES

The plan for financing the construction, maintenance, and operation of the dam, reservoir, power plant, and incidental structures is distinctly different from the financial plan for the All-American Canal. It is not a plan which comes within the scope of the Reclamation Law and is, therefore, outlined in detail in the Boulder Canyon Project Act. The act provided for a fund, known as the Colorado River Dam Fund, the resources of which are to be used only for carrying out the provisions of the act. All revenues received under the act are to be paid into the fund under the direction of the Secretary of the Interior. By using a special fund of this kind, the receipts and expenditures of the Boulder Canyon Project are kept separate from those of any other project and from the general funds of the federal government. At all times, therefore, from the beginning of construction throughout the life of the project, an account of the receipts and expenditures should be readily available and the financial condition easily determinable.

Revenues of the project will come chiefly from the sale of power (under one of the three alternatives) and from the sale of stored water. Some revenue is derived from the small fees charged tourists who visit the dam, and in the future another comparatively small source of income may be obtained if the territory near the reservoir becomes the site of a pleasure resort. The two sources of revenue upon which the financial success of the project depends, however, are power and water; and, of these two, power is by far the more important. These revenues will be paid into the Colorado River Dam Fund; and on June 30, at the end of each fiscal year when all expenditures including construction, operation, maintenance, and interest have been accounted for, any excess revenues will be covered into the Treasury of the United States as repayments on advances from the federal government. Thus the excess revenues are to be used first to cover the scheduled payments to amortize the advances made to the project from the federal treasury. The act also provided that the excess revenues, if any, above the amortization requirements will be divided on the basis of 62½ per cent for the repayment of the federal government advances for flood control, 18¾ per cent for the state of Arizona, and 18¾ per cent for the state of Nevada. The act appears to be somewhat inconsistent in these provisions, since in Section 2*e* all excess revenues were allocated to the federal treasury, while in Section 4*b*, 37½ per cent of the excess revenues above amortization requirements were given to the states of Arizona and Nevada. Undoubtedly the intent was to provide a source of revenue from the project for Arizona and Nevada, since the 62½ per cent figure for repayment of advances for flood control was stated in Section 2*b* of the act and the two 18¾ per cent figures were given definitely in Section 4*b*.[31] After repayment with interest has been made of all money advances by the United States, the excess revenues not otherwise allocated are to be kept in a separate fund to be expended within the Colorado River Basin as prescribed some time in the future by Congress. After the cost of the project has been repaid, therefore, the net revenues are to be spent for the advancement of the Colorado River Basin and are not to be subject to appropriation for the benefit of other parts of the country. In all probability this fund will provide the financial basis for future projects which will carry on and expand the development of this part of the arid West.

To carry out the provisions of the Boulder Canyon Project Act, the Colorado River Dam Fund will receive from time to time appropriations from the federal treasury. The Secretary of the Treasury was authorized to advance to the fund such sums as the Secretary of the Interior might deem necessary to carry on the work; but the total amount of these advances was limited to $165,000,000. Although the financial plan of the All-American Canal differs from the financial plan of the rest of the project, all advances made by the federal govern-

[31] *United States Statutes at Large*, Vol. XLV, Part 1, pp. 1057–59.

ment for the building of the All-American Canal are to be paid into the Colorado River Dam Fund and are to be included in the $165,000,000 maximum. Therefore, the construction cost of the entire project to the United States, including the dam, reservoir, power plant, machinery, All-American Canal, appurtenant structures, and interest during construction, cannot exceed $165,000,000. It should be noted, however, that this money is available only for the payment of construction costs and for the payment of interest. Expenses for operation and maintenance are to be paid from specific appropriations therefor and are not to be confused with the appropriations made to build the project. The interest charge was set at 4 per cent per annum and is to apply to all sums advanced to the fund, excepting that advances for the All-American Canal are relieved of the interest burden under the provisions of the Reclamation Law. For all other sums advanced and not repaid, the Secretary of the Treasury must compute the accrued interest as of June 30 of each year and charge the Colorado River Dam Fund with the amount due. If the fund is insufficient to meet the interest payment, the Secretary of the Treasury may defer the payment, or any part of it, to a future date; but the amount so deferred must be charged interest at 4 per cent until payment is made. Thus some leeway is given in meeting the financial provisions of the act, and inability to meet interest payments does not mean necessarily that the project has failed to live up to the requirements of the act. Interest, of course, is a necessary element in every sound financial plan. To have omitted the very reasonable interest charges required on advances to the Boulder Canyon Project would have distorted greatly the cost estimates, and as a result the error would have been reflected in the power and water rates charged to secure sufficient revenues to repay to the government the cost of the project.

Of the $165,000,000 which may be appropriated for the project, $25,000,000 was allocated to flood control. The $25,000,000 is not to be repaid to the federal government according to the same method devised for the rest of the appropriation but is to be repaid out of 62½ per cent of the revenues, if any, in excess of the amounts necessary to meet periodical payments during the period of amortization. In other words all other payments take precedence over the repayment of the flood-control allocation during the fifty-year amortization period. It was provided further that if the $25,000,000 is not repaid during the amortization period according to the above arrangement, it is to be repaid from 62½ per cent of all revenues after the amortization period has ended. This means that the flood-control allocation must be repaid but that it need not be repaid during the amortization period unless the funds are available. Thus the burden assumed by the project to repay its cost to the federal government within a certain time may be relieved to the extent that the repayment of the $25,000,000 may be deferred to a future period. The reason for this leniency as far as the flood allocation is concerned is that the federal government recognized a certain obligation resting upon it to provide flood control for the lower reaches of the Colorado River, and did not wish to place too heavy a burden on the project in meeting this obligation. The fourth Swing-Johnson bill, which became the Boulder Canyon Project Act, had originally provided for a maximum appropriation of $125,000,000, but this amount was increased by amendment to provide for the $25,000,000 flood-control allocation and for a higher dam than the previous plans had called for.[32] All other advances made by the federal government for construction plus accrued interest must be repaid according to the amortization schedule which is extended over a period of 50 years from the date of completion of the work. Expenditures for the All-American Canal, of course, including expenditures for operation and maintenance, are to be repaid in the manner provided by the Reclamation Law.

In order to meet the repayment obligations imposed by the act, the Secretary of the Interior was authorized to execute contracts for the sale of power and for the sale of stored water which would, in his judgment, insure revenues sufficient to operate and maintain the works provided for in the act and to repay their cost to the federal government within fifty years of the date of

[32] *Congressional Record,* 70th Congress, 2d Session, Vol. LXX, Part 1 (1928), p. 399.

completion of the works. The act provided that these contracts had to be executed before any money could be appropriated for the construction of the project; hence the plan for repayment to the federal government had to be worked out in detail before construction could begin.[33] The contracts respecting water for irrigation and domestic uses were to be for permanent service; but the contracts for electric energy were to be limited in duration to a maximum of fifty years. Water is a necessity of life, and to cancel or to refuse to renew a water contract might mean the death of a community or the abandonment of lands which had been developed on the basis of contractual agreements for water. Therefore it is only fair that water contracts should be on a permanent basis. Electric energy is not an absolute necessity of life, however, and there is no reason why the federal government should by contract permanently give up its control over the power generated at the dam. The contracts for electric energy may be renewed subject to the laws and regulations existing at the renewal date; and, in the event that renewal is refused, the government is obligated to compensate the holders of the contracts for property dependent for its value upon the continuation of the contracts. Therefore, refusal to renew a contract after expiration could not be construed as confiscation of property, especially since the fifty-year period should give the contract holder ample time to reap the benefits of the contract and to amortize his investment.

In the case of either the water contracts or the power contracts, the Secretary of the Interior was authorized to prescribe general and uniform regulations for the awarding of contracts. The contracts were to be made with responsible applicants who would pay the price fixed by the Secretary of the Interior with a view to meeting the revenue requirements of the act. If there should be conflicting applications, preferences were to be given in accordance with the Federal Water Power Act. That is, states and municipalities were to be given preference over other applicants if their plans for using the project were equally as well

[33] *United States Statutes at Large,* Vol. XLV, Part 1, Sec. 17, p. 1065, provided that claims of the United States arising out of any contract authorized by the act should have priority over all others, secured or unsecured.

adapted to the public interest. As between other applicants, preference was to be given to the applicant presenting the best plan for development.[34] The Boulder Canyon Project Act qualified this schedule of preferences still further by stating that the first preference should be given to a state, should one apply, and that the states of Arizona, California, and Nevada should be given equal opportunity as such applicants. This last provision does not give the lower-basin states preference over other states but merely recognizes the fact that, because of their location, these three states are the most likely states to apply. The act provided that a state applying for electric power must notify the Secretary of the Interior and must enter into a contract within the following six months on the same terms and conditions as might be provided in other similar contracts. An applicant state was not to lose its preference merely because the bond issue necessary to build the works to utilize the power had not been authorized or marketed; but the preference was to be lost if, after a reasonable length of time, the state had not prepared itself to utilize the power beneficially. These provisions were included to prevent states from reserving electric energy which they could not use, thus impeding the fullest and most economic use of the project.

To carry the power to the various markets where it will be used, transmission lines must be built by the successful applicants for the power. The building of these lines will involve an expense which will be worth while only to an applicant who can use and who has contracted for a large amount of power. Both Boulder and Black canyons are so far from power markets that the possible applicants for Boulder Canyon Project power would be reduced to a few very large users if arrangements for joint use of transmission lines were not made. Therefore, a provision was included in the act which gave the Secretary of the Interior the authority to require a large contractor to permit a small contractor to share his transmission lines. That is, an agency receiving a contract for 100,000 firm horsepower may, if economic and engineering conditions warrant it, be required to share the use of its main transmission lines with an agency receiving a

[34] *United States Statutes at Large,* Vol. XLI, Part 1, chap. 1063, Sec. 7, p. 1067.

contract for not over 25,000 firm horsepower. The large contractor cannot be called upon to give up more than one-fourth of the capacity of his transmission line, and the small contractor must bear a reasonable share of the cost of construction, operation, and maintenance of the line. This provision for sharing transmission lines had to be made to secure a wide use of the power and the satisfaction of the most urgent needs. In addition, the waste of money through duplication of facilities caused by the construction of parallel lines will be avoided in cases where one line can give adequate service. In this way all agencies, large and small, which are within reach of the transmission lines built from the dam may receive electric power according to their needs, and the widest and most economic distribution of the power benefits will be achieved.

COMPACTS

The Boulder Canyon Project Act ratified the Colorado River Compact on behalf of the United States on either the seven-state or the six-state basis.[35] The act made the compact binding upon approval by the seven states or upon approval by six of the seven states, including California, and provided for a waiver of Article XI of the compact requiring unanimous ratification. This provision made possible the final approval of the compact without Arizona's consent.

The act also authorized the states of the Colorado River Basin to enter into further compacts or agreements supplemental to the Colorado River Compact and consistent with the Boulder Canyon Project Act for a comprehensive plan for the development of the Colorado River. These agreements may provide for the construction of dams, reservoirs, diversion works, power plants, etc., to secure flood control, reclamation of lands, and improvement of navigation with the development of hydroelectric power to finance the projects. Interstate commissions and other instrumentalities may be formed to achieve these purposes, but in every case a representative of the United States appointed by the President is to participate in the negotiations and to make a report to Congress of the proceedings and the provisions of

[35] *United States Statutes at Large,* Vol. XLV, Part 1, Sec. 13*a*, p. 1064.

any compact entered into. Thus the federal government has recognized the fact that future compacts and future construction will be necessary to secure the full development of the Colorado River. The seven states were given the authority to enter the necessary agreements, but the negotiations are always to be under federal surveillance. Evidently the federal government felt that the development of the Colorado River is too important to the national interest to be left entirely to the activity of the seven states without federal control and co-ordination. The act provided further that none of these compacts is to be considered binding on a state until it has been approved by the legislature of that state and by the Congress of the United States. Thus the sovereign power of a state was recognized in that no state can be forced into a compact against its will; but, at the same time, state enthusiasm for an unwarranted development may be curbed by the veto of the federal government.

At the time the Boulder Canyon Project Act was voted upon, there was still some hope that Arizona, California, and Nevada would reach an agreement concerning the division of the waters of the lower Colorado River and that Arizona would then ratify the Colorado River Compact. In order to facilitate such an agreement, the act gave the three states the authority to enter a compact under certain conditions.[36] These conditions were that: (1) Of the 7,500,000 acre-feet annually apportioned to the lower basin by the Colorado River Compact, Nevada was to receive 300,000 acre-feet and Arizona 2,800,000 acre-feet for exclusive beneficial consumptive use in perpetuity; this, of course, left 4,400,000 acre-feet for California. (2) Arizona was to have the privilege to use annually one-half of the surplus waters unapportioned by the Colorado River Compact; that is, Arizona was to be allowed the use of one-half of the water found to exist in the stream after the 7,500,000 acre-feet allocated to the upper basin and the 7,500,000 acre-feet allocated to the lower basin had been deducted. (3) Arizona was to have the exclusive beneficial consumptive use of the Gila River and its tributaries within the boundaries of the state. This provision increased greatly Arizona's allocation of water, since

[36] *United States Statutes at Large,* Vol. XLV, Part 1, Sec. 4*a*, p. 1059.

the Gila is an important stream; but it was not entirely consistent with the Colorado River Compact, which treated the waters of the Colorado River Basin as a system and divided the total accordingly. Therefore the waters of all tributaries were included in the allocations, and the waters of the Gila were made a part of the 7,500,000 acre-feet allocated annually to the lower basin. As pointed out before, however, the waters of the tributaries were not included in the total 75,000,000 acre-feet each ten years upon which the lower basin had first call; thus, with storage, the lower basin would have enough main stream water to fulfill the provisions of the suggested agreement and to reserve the waters of the Gila for Arizona besides. (4) The waters of the Gila were not to be used to meet the requirements of a treaty under which the United States might grant an allowance of water to Mexico. To be consistent with the compact, such an allowance would have to come first from the surplus waters as defined by the compact; and, if that were not sufficient, California and Arizona would each have to supply one-half of the deficiency from their respective shares of the main stream water of the Colorado River. (5) Arizona, California, and Nevada were mutually to agree that none of the three would withhold water that could not reasonably be used. (6) The provisions of the three-state agreement were to be subject in all particulars to the Colorado River Compact. And (7) the agreement was to take effect only upon the ratification of the Colorado River Compact by all three states.

The fact that so much attention was given in the Boulder Canyon Project Act to a possible tri-state agreement indicated that the federal government was anxious to have complete harmony among the states of the lower basin before the project was begun. Yet it also indicated that the federal government had become weary of the arguments between the states (especially between Arizona and California), and that after years of waiting it was finally suggesting its own solution to the problem. Neither acceptance nor refusal of the agreement would affect the passage of the bill. The United States had decided that a project of such vital concern to the lower-Colorado region and to the general interest of the entire country should be delayed no longer by the quarrels between the states, and this last opportunity was offered to settle all differences. Naturally the government considered the proposed compact fair to all parties concerned; but Arizona still felt that she deserved more consideration, and the agreement was not consummated.

RIGHTS TO WATER AND ELECTRIC POWER

The act specifically provided that all rights to the waters of the Colorado River should be subject to and controlled by the Colorado River Compact.[37] No attempt was made to change the laws of the states; but, where state regulations concerning the appropriation of water within their borders did not agree with the compact or other interstate agreement, the state right to enforce such regulations was modified accordingly. The act provided that all users of the project, whether political subdivisions[38] or private parties, should observe the regulations of the compact, and that even in matters of construction, management, and operation of the project the policies outlined in the compact were the ones to be followed. Therefore the entire act was based upon the Colorado River Compact and the compact, rather than the Boulder Canyon Project Act, is to be the authority in case there is a conflict between the two concerning the storage and diversion of water or the generation of power or the development of any other phase of the project. The operation and use of the project were also made subject to any compact that might be entered into by Arizona, California, and Nevada before January 1, 1929, or any compact among them approved by Congress after that date, provided that in the latter case the compact should be subject to all contracts made by the Secretary of the Interior prior to the date of such approval. Such an agreement would be subject, of course, to the terms of the Colorado River Compact, which is still, in the last analysis, the highest authority. All permits, contracts, leases, or other privileges granted by the United States, or under its authority, for any or all of the vari-

[37] *United States Statutes at Large*, Vol. XLV, Part 1, Sec. 13*b*, p. 1064.

[38] *United States Statutes at Large*, Vol. XLV, Part 1, Sec. 12, p. 1064, defines "political subdivision" as any state, irrigation or other district, municipality, or other governmental organization.

ous uses of the Colorado River are to contain the express condition that the holders of these privileges shall observe the regulations of the Colorado River Compact. Whether this condition is specifically mentioned in the permits, contracts, leases, etc., or not, it shall be considered as a matter of law to be a part of all such instruments; and, as a legal part of these instruments, it is available to the seven Colorado River Basin states and to the users of the Colorado River water by way of suit, defense, or otherwise in any litigation respecting the waters of the Colorado River or its tributaries.

Thus all rights for water, or for power, or any other uses of the project are dependent upon the provisions of the Colorado River Compact. The importance of this document to the Boulder Canyon Project and the future development of the entire Colorado River Basin cannot be overestimated. The rights of Mexican users, as well as American users, were provided for in the compact. The Boulder Canyon Project Act stated that nothing in the act should be construed as a denial or recognition of Mexican rights, if any; but if the United States should grant to Mexico by treaty a definite yearly quota of water from the Colorado, the obligations of the treaty undoubtedly would be met according to the method outlined by the compact. Therefore the Colorado River Compact must always be considered wherever rights to the use of the Colorado River are concerned, no matter when or how these rights were acquired or by what authority they were granted.

INVESTIGATIONS

The Boulder Canyon Project was the first big step taken by the federal government to develop the resources of the Colorado River Basin; but it was not expected to be the last. Undoubtedly when economic conditions warrant the expansion the federal government will develop future projects in the Colorado region. In order that this development may take place in the most economic and effective way according to a well-devised plan, the Boulder Canyon Project Act authorized the appropriation of $250,000 from the Colorado River Dam Fund to investigate the feasibility of irrigation, power, and other projects in the

Colorado River Basin.[39] It is obvious that a comprehensive plan for the control, improvement, and utilization of the resources of the Colorado River system will be an aid in administering present and future projects in the Colorado River Basin as a unit, and will also permit the most advantageous development with the least waste in resources. Not all of the work of devising this plan was left to the Secretary of the Interior as representative of the United States, however. Commissioners from the ratifying states were given the right to act in an advisory capacity to the Secretary of the Interior and to co-operate with him in formulating the plan of development. This provision granted the states the right to express their opinions and to use their influence as to these matters which are of such vital importance to them.

The Secretary of the Interior was also authorized to make such studies, surveys, and investigations as may be necessary to determine the lands in the state of Arizona which should be embraced within the boundaries of the Parker-Gila Valley Reclamation Project. In general, this project is taking into consideration lands north and south of the Gila River, in the Cibola Valley, in the Parker Valley (Colorado River Indian Reservation), and on the Parker Mesa, all of which are in Arizona. In addition, the project may include the reclamation of California lands in the Palo Verde Valley, on the Palo Verde Mesa, in the Chucawalla Valley, and in the Parker Valley; but the Boulder Canyon Project Act merely authorized the Secretary of the Interior to determine the most feasible methods of irrigating the Arizona lands and to report to Congress his findings and recommendations. Although this provision directed attention to the possibility of the reclamation of lands in Arizona, it appears to be a superfluous addition to the legislation, since investigations of the entire basin were substantially provided for in other sections of the act and in the Colorado River Compact.

CONDITIONS

The favorable vote in Congress on the Boulder Canyon Project Act and its approval by President Coolidge did not make the

[39] *United States Statutes at Large*, Vol. XLV, Part 1, Sec. 15, p. 1065.

act effective or mark the beginning of the construction work. Certain conditions provided for in the act had to be fulfilled before work on the project could be begun. In the first place, it was provided that the act should not become effective until the Colorado River Compact had been ratified on either the seven-state or the six-state basis.[40] To give all seven states sufficient time to ratify the compact, a six-state ratification would not be considered binding until six months after the passage of the act. Unless Arizona signed the compact, the act could not possibly become effective until at least six months had elapsed.

In the second place, the act could not become effective until the state of California, by act of its legislature, had agreed irrevocably and unconditionally with the United States and for the benefit of the other Colorado River Basin states to limit her aggregate annual consumptive use of water from the Colorado River to a maximum of 4,400,000 acre-feet of the waters apportioned to the lower basin by the Colorado River Compact, plus one-half of the surplus waters unapportioned by the Compact. The state of California enacted the required legislation on March 4, 1929, and with the ratification of the Colorado River Compact on March 6, 1929, by Utah as the sixth state, the first two conditions of the act became satisfied with only the lapse of the six months' period from the date of passage of the act necessary to complete fulfillment.

On June 25, 1929, President Hoover declared the conditions of the act to be fulfilled and proclaimed the act effective as of that date.[41] One more step remained to be taken, however. The act provided that no money could be appropriated for construction of the project until the Secretary of the Interior had provided, by contract, revenues adequate to pay operation and maintenance expenses and to meet repayment requirements. No provision of this nature had ever been included in previous Congressional legislation.[42] The act became effective with Presi-

[40] *United States Statutes at Large,* Vol. XLV, Part 1, Sec. 4*a*, p. 1058.

[41] *Ibid.,* Vol. XLVI, Part 2, p. 3000.

[42] House Hearing before the Committee on Rules, 70th Congress, 1st Session, *Boulder Dam* (1929), Part 2, Statement of Hon. Philip D. Swing, a Representative in Congress from the State of California, p. 81.

dent Hoover's proclamation, but it could not lead to tangible results until the Secretary of the Interior had secured sound and binding contracts which would guarantee the return to the federal government of every dollar expended by it plus an interest charge of 4 per cent on the greater part of the sum advanced. In April 1930 two contracts for power and one for the delivery of water were executed which fulfilled the final condition of the act. On July 3, 1930, the Second Deficiency Appropriation Bill for 1930, containing an item of $10,660,000 to start work on the Boulder Canyon Project, was approved;[43] and on July 5 actual work on the project was begun.

ARGUMENTS OF THE OPPOSITION

Although the Boulder Canyon Project Act had received the support of Congress and the approval of the President of the United States, the opponents of the project continued their fight against it in an attempt to prevent the actual construction of the dam, power plant, and canal. This fight was led by the state of Arizona; but the project was so tremendous and the issues so vital that the progress of the development received considerable attention in all parts of the country. Many people who were neither directly interested in the project nor in sympathy with Arizona's point of view opposed the development on the grounds that it was unsound, unwarranted, or in violation of certain established principles of American business and government. They allied themselves with Arizona in an attempt to prevent the establishment of a precedent which appeared to them to be both dangerous and unnecessary. Even after construction had started the opponents of the project continued their verbal battle against it. They stated that the project as planned was neither a desirable nor a safe type of construction from an engineering standpoint, and they maintained that the project was economically unsound since it was not needed for flood control or to supply water for domestic and agricultural uses and for the generation of electric power. These engineering and economic arguments will be discussed in detail in later chapters. But here the analysis

[43] *United States Statutes at Large*, Vol. XLVI, Part 1, p. 877.

will be confined to four other objections made by the opposition: (1) It was argued that the development should be left to state initiative and not placed under federal control. (2) It was stated that the project would permit the federal government to invade the field of private industry by entering the power business. (3) It was held that a formal agreement with Mexico should be concluded to clarify the international situation before work was begun. (4) It was maintained that the entire project was illegal and unconstitutional.

STATE VERSUS FEDERAL CONTROL

Many of those who opposed the Boulder Canyon Project believed that such a development should be left to state initiative. They pointed out that the states of Arizona, California, and Nevada would receive the greatest benefits of the project, and they argued that the whole development was really more of a local problem than a national affair. They seemed to fear that national control in this case would establish a precedent for federal bureaucratic control over vital industrial resources of the states.[44] They said that the project was a challenge to the right of all states to develop the economic resources within their borders. The opponents claimed that the project was a huge business enterprise which the states within whose borders the resources lie could not tax or control, and that the whole future development of the arid Southwestern states might now be left entirely in the hands of the Bureau of Reclamation of the Department of the Interior. The precedent established was criticized as being especially dangerous in view of the fact that the Colorado River Compact which would have maintained the right of the states to consent or object to similar federal invasions had not been signed by Arizona, one of the states most vitally concerned. If the power resources of Arizona could be taken from her against her will, it was argued that resources such as coal or oil might be seized in other states and placed under federal control.

In addition to the argument that the Boulder Canyon Project

[44] E. O. Leatherwood, "My Objections to the Boulder Dam Project," *The Annals of the American Academy of Political and Social Science*, Vol. CXXXV, No. 224 (January 1928), pp. 133, 135, 138.

would rob the states of their right to control the use and distribution of water and other resources within their borders, the development was attacked on the ground that the action of the federal government in usurping the control of resources belonging to Arizona and Nevada was for practically the sole and exclusive benefit of California. The opponents of the project pointed out that the most vital need for flood protection was in the Imperial Valley of southern California. It was the state of California which had pushed the reclamation of its desert lands to the limit of the supply of water available and which would undoubtedly be the first to expand its cultivated acreage when the regulated flow of the river created a larger and more dependable supply. The All-American Canal was to be built entirely within California territory for the benefit of California lands. By far the greatest part of the Colorado River water used for potable purposes would go to certain cities of southern California which had outgrown their local sources of supply. And, finally, the power generated at the Black Canyon dam would be used chiefly in California, although the power plant was not even located in that state. The opponents claimed that it was ridiculous to place such a project under national control. They intimated that the proper procedure would be to permit the states or private interests to carry out the development according to the provisions of some agreement between the states and the rules of the Federal Power Commission. Naturally the state of California or California interests would be required to bear the greatest part of the expense of the development, since inhabitants of California would receive the greatest benefits from the project. The project should be managed as a business venture, and, as in a business venture, the charges should be made according to the services rendered.

The supporters of the Boulder Canyon Project could not deny with reason that the project had established a precedent as far as the problems of flood control, water storage, and hydroelectric power, and their relationship with federal and state governments were concerned. It was true that for years the federal government had been active in carrying out various reclamation projects; but none of these previous projects could compare in size or in importance with the Boulder Canyon development. Never before had federal activity in the field of reclamation involved such far-reaching decisions concerning the relative merits of state and federal control and the limits of their influence. The supporters of the Boulder Canyon Project were forced to admit that the project had pointed toward increased federal influence in the fields of reclamation and hydroelectric power, and it seemed reasonable to expect that this increased activity would not be confined to those two fields alone. In the writer's opinion, however, the mere fact that the federal government had invaded, or might invade, certain spheres previously occupied only by state or private activity was beside the point. The important question was to determine how the resources of the country could be exploited most economically and used most advantageously. If the national government proved to be the most efficient agent for the work, federal activity should be encouraged. It would be foolish to cling to state or private control alone when the good of the people could be served best by the centrally directed activity of the federal government.

As far as the Boulder Canyon Project was concerned, the advantages of federal control were so great that the advisability of federal construction, supervision, and operation was hardly open to question. The federal government was the only agency in a position to solve the international problems involved. Individual states or groups of states did not have the constitutional authority to enter into agreements with foreign governments without the consent of Congress.[45] Therefore an agreement with Mexico to divide the waters of the lower Colorado River could be binding only if it were negotiated by the federal government of the United States. Negotiations between the states and Mexico alone, no matter how enlightening, could not possibly be fruitful without federal approval and federal action. Even if the states had the authority to enter into agreements with foreign nations, it would be unwise to allow them to negotiate with Mexico in this case. The states are not in as strong a position

[45] *The Constitution of the United States*, Art. 3, Sec. 10, states, "No State shall without the consent of Congress, ... enter into any agreement or compact with another State or with a foreign power, ..."

as the federal government in dealing with foreign nations, and it was reasonable to expect that an agreement more advantageous to the United States could be reached through federal than through state negotiations.

Without federal supervision and control, the Boulder Canyon Project could have proceeded only upon the basis of a co-operative agreement between at least the three lower-basin states, since the Colorado River is an interstate as well as an international stream. On the supposition that the upper-basin states would not object to the development, Arizona and Nevada would still have to agree to the building of the dam on their territory, and California would have to agree to buy a certain amount of power and to limit her use of water. The long negotiations involved in attempting to secure the ratification of the Colorado River Compact have proved how difficult it is for the various states to iron out their differences; and even if an agreement were reached it would be constitutional only when approved by the federal government. In view of these facts, therefore, it was obvious that the federal government was a necessary consideration to the project under any circumstances.

It was possible, but not probable, that the various states would reach a satisfactory agreement to carry out the project. But federal control would still be more advisable than state control. The entire watershed is a unit, and its use and development could not be limited physically by state or international lines. Any proposed development at any point along the course of the river would have to be considered with regard to its effect upon the entire river. The jurisdiction of a single state was not broad enough to cope with such a situation. The general supervision by all seven Colorado River Basin states also would not lead to a satisfactory solution, since each state would be more interested in its own development than in the most desirable, unified development of the entire basin. The advantages of centrally directed activity were so great in this case that only those who objected to the general principles of the extension of federal control could reasonably oppose the project on this ground alone. In addition, the federal government had had many years of experience, through the activities of the Bureau of Reclamation of the Department of the Interior, in the building of dams for reclamation projects. A project of the size of the Boulder Canyon development had never been attempted before, and many new problems were involved. The facilities possessed by the Bureau of Reclamation, however, made that organization the most logical agency for direct supervision of the project. To have placed the planning and construction of such a tremendous project into the hands of a comparatively untried state organization would have been a serious, and perhaps fatal, blunder.

The federal government had still other reasons to be vitally interested in this development on the Colorado River. The great dam is situated on United States government property, and most of the land submerged by the reservoir was also a part of the public domain. Under state development the use of this land could have been secured only through Congressional act, and undoubtedly federal supervision would have been required to protect the federal interests involved. The Yuma project, a federal reclamation development, was directly affected by the Boulder Canyon Project; and this fact immediately aroused the interest of the national government. Flood-control and reclamation projects had long been accepted as activities within the scope of federal operation, and the possibilities of power development brought that part of the project under the supervision of the Federal Power Commission. With all of these interests involved, it was obvious from the beginning that the federal government would play a leading part in the Boulder Canyon development.

Aside from the vital interests of the national government in the Colorado River and the necessity for federal control during the general development of the project, federal supervision was also essential to the most efficient operation of the dam and power plant. There are various conflicting uses of the project, and such problems as the quantities of water that should be stored for future irrigation needs and the quantities that should be permitted to flow around the dam to generate power had to be solved. The federal government was the proper and logical agency to decide these questions, especially when one state might want water stored for future irrigation while another might want a steadier flow to generate more power. No one state had the authority to deal with such broad questions, and an agreement between states as to the policies to be followed would take months or years of negotiations and would probably be too inflexible for practical application.

Federal control was especially important where operation of the power plant was concerned. There is only one natural site for a power plant, at the base of the dam in Black Canyon. The canyon is very deep and very narrow, and the successful private bidder for the one site available would have had a tremendous advantage over all his competitors. It would have cost an immense sum of money for another agency to blast down the canyon walls and to prepare a similar site.[46] Therefore it was practical to have only one unified power plant, and, unless a state, municipal, or private monopoly was to be fostered, the control of the power plant would have to be kept in the hands of the federal government. Under central control the power generated could be distributed equitably to the state, municipal, and private applicants, who would probably receive much fairer treatment from the federal government than from any other agency controlling it. Controversies between applicants could be settled without long delays in adjustment, and the power could be distributed according to a schedule to fill the greatest social need rather than to promote the unwarranted development of some community or private enterprise.

It was an undisputed fact that California would receive the greatest immediate benefits of the Boulder Canyon Project. The danger that the Imperial Valley might be permanently inundated would no longer exist, and the regulated flow of the river and the building of the All-American Canal would make possible the reclamation of wide areas of California desert lands. Contracts had been signed to provide a number of cities of southern California with water diverted from the Colorado River, and most of the power generated was to be used in California. Yet these facts did not prove the opponents' arguments that the en-

[46] House Hearings before the Committee on Irrigation and Reclamation, 70th Congress, 1st Session, *Protection and Development of the Lower Colorado River,* Part 1 (1928), Statement of Mr. Swing, p. 14.

tire project was for all practical purposes a California project and that it was undertaken to rob Arizona and Nevada for California's benefit. The great dam was built primarily as a flood-control and water-storage project for the lower basin of the Colorado River, and not for the exclusive benefit of any one state. Arizona and Nevada were both to receive an income from the power generated at the dam. Arizona's future agricultural development would be greatly enhanced by the regulated flow of the river, and the projects at Yuma and other points on the Arizona side of the river would no longer be subject to destructive floods. As it happened, California was in a position to take immediate advantage of the benefits of the Boulder Canyon development, while many of the project's benefits to Arizona and Nevada would be necessarily postponed to a future period when the development of those states permitted their realization.

California does not have financial resources comparable to those of the federal government, and undoubtedly the state would not have attempted a project which would require the outlays necessary to construct the dam at Black Canyon, even if Arizona and Nevada had given their support and co-operation. Still the fact that California did not build the dam and other structures and that federal funds were used to meet construction expenses did not warrant the conclusion that California would receive the benefits of the project as an outright gift from the federal government. Actually the cost would be borne chiefly by the people of southern California. According to the financial plan outlined in the Boulder Canyon Project Act, the federal government was to be reimbursed for its expenditures on the dam and appurtenant works through the sale of power and the sale of stored water. Most of the power generated and a great part of the water stored would be sold to California users, who would repay to the federal government all but a minor share of the cost of this construction. The All-American Canal feature of the project was really a California development, and its cost was to be repaid by the California lands benefited under the provisions of the Reclamation Law. Therefore the cost of the Boulder Canyon structures, which was originally imposed upon the federal government, would be largely shifted to those receiving the benefits of the project, and California interests would be forced to assume the greater part of this burden. The argument of the opponents that the development was for the sole and exclusive advantage of California not only overlooked the benefits to be realized by other states but also omitted the important consideration that eventually California interests would have to repay the federal government for most of the disbursements made.

GOVERNMENT IN THE POWER BUSINESS

A second objection to the Boulder Canyon Project was based on the premise that it would permit the federal government to invade the field of private industry by entering the power business. The development was branded as a gigantic power scheme masquerading as a flood-control and water-storage project. It was claimed that the primary purpose of the project was to establish the federal government in the power business and that the flood-control and water-storage aspects were merely incidental. On the assumption that these statements were true, the opponents criticized the project as being a great stride away from the sound, well-established, and traditional policy that the government should not compete with private industry. They feared that such action would establish the dangerous precedent that the federal government may undertake any business enterprise under the guise of carrying on a legitimate governmental function when the relationship between the two is based solely upon the hope that the possible profits of the business enterprise will cover the expense of the legitimate function. The existence of that fear aroused the active opposition of the power interests and a number of business organizations to the project, and this opposition was one reason that the Boulder Canyon Project Act met so many obstacles in the course of its passage through Congress. The opponents admitted that the federal government had already built some fifteen small power plants in connection with reclamation projects; but they pointed out that the total cost of these plants was slightly less than $1,500,000 and that the average cost was less than $100,000.[47] These figures they considered

[47] Leatherwood, *op. cit.*, pp. 133, 146.

so small in comparison with the $38,200,000 power development at Black Canyon that the earlier power developments could not be construed as a precedent giving the federal government the right to enter the power industry on such a gigantic scale. In addition, it was argued, the huge size and great cost of the power plant proved definitely that the project was primarily a power project and not a reclamation project with power generated as a by-product.

In the writer's opinion this argument of the opposition was based upon the erroneous assumption that the Boulder Canyon Project is primarily a power project. From the very beginning of the development the project was planned as a reclamation and flood-control project. The power features were a later addition to the plans for Colorado River development and were not a part of the original Boulder Canyon Project bill when it was first submitted to Congress. The power features were incorporated in the bill at the request of the Secretary of the Interior and not at the insistence of the proponents of the legislation.[48] In the Boulder Canyon Project Act the generation of electric energy was given as the fifth and last purpose of the project and as the third and last preferred use. River regulation, navigation, flood control, and water storage were all given preference over the generation of power. It is quite true that the sale of power generated at the dam is expected to pay for the project within a period of fifty years; but the legislation enacted merely authorized the Secretary of the Interior to build the plant if he deemed it necessary or advisable on the basis of engineering or economic considerations.[49] Therefore the power plant was not classified in the act as an essential or necessary feature of the project. It should be noted, however, that undoubtedly the federal government intended to exploit the power possibilities of the project if the legislation were passed, and thus the actual importance of the power feature was far greater than its classification would indicate.

[48] House Hearings before the Committee on Irrigation and Reclamation, 70th Congress, 1st Session, *Protection and Development of the Lower Colorado River* (1928), Statement of Mr. Swing, p. 14.

[49] *United States Statutes at Large,* Vol. XLV, Part 1, p. 1057.

From the standpoint of cost of construction it is true that the amount of money required to build the power plant was very large. Compared with this outlay, the costs of other government power projects seem insignificant. Still the $38,200,000 power plant is only a part of a $165,000,000 project, and this total cost figure may be increased to $365,000,000 if the related $200,000,000 Colorado River Aqueduct Project is included. These relative costs show that the power plant does not occupy a position of overshadowing importance as far as the Boulder Canyon Project as a whole is concerned. When consideration is given to the necessity of (1) preventing the destruction of such communities as those in the Imperial Valley by floods or (2) assuring an adequate municipal water supply for more than 2,000,000 people in Los Angeles County or (3) providing sufficient water to irrigate large areas of arid land, the power features of the project seem much less important than the arguments of the opponents of the project would indicate.

Power is often a by-product of government works built for flood control, reclamation, and other purposes regularly included within the scope of federal activity. Such incidental power production does not mean that the government has entered the field of power generation and distribution as a business.[50] In the case of the Boulder Canyon Project, the activities of the federal government are limited to the sale of power at the switchboard or to the leasing of the power plant and the equipment. The Boulder Canyon Project Act provided that contracts to dispose of the electric energy generated are to be executed between the federal government and the state, municipal, and private power interests. The government may generate the power at the dam, but the power must be sold to the agencies which have contracted for it. No provision has been made for the transmission of power by the federal government to the power markets, and the act provided that the contracting agencies must build their own transmission lines. The distribution of power to the ultimate consumers, therefore, was left entirely within the hands of existing power interests. Under such a situation, the statement that

[50] Herbert Hoover, "The Case against Government Ownership of Utilities," *Engineering News-Record,* Vol. XCIII, No. 16 (October 16, 1924), p. 627.

the government had entered the power business is farfetched. The federal government is not competing with existing power interests but is co-operating with them to dispose of the power to the mutual benefit of all parties concerned. The power interests were not forced to enter into contracts to buy the power. The fact that the federal government experienced no difficulty in finding applicants for the power under the terms offered indicates that the contracting agencies felt that the Boulder Canyon Project had resulted in a situation which would prove profitable to their own businesses.

THE MEXICAN SITUATION

The question whether or not the federal government had entered the power business was of great interest to the power industry and to those business organizations which opposed on general principles the possible creation of a precedent which might establish the government's right to compete with private business. To the general public, however, this argument was not of such vital importance. The power interests had received unfavorable publicity concerning the monopolistic aspects of their business, and public opinion had changed to such an extent that the concept of government control of certain industries was no longer regarded as a violation of the freedom and rights of the people. Therefore the opponents of the Boulder Canyon Project developed another argument, which not only seemed logical to many observers but which appealed to their patriotic emotions. They based this objection on the premise that Mexico would share in the benefits of the project without contributing anything to defray the costs. It was pointed out that the Republic of Mexico is entitled to half of the water which passes through the Imperial Canal, and that the acreage of Colorado River delta land under cultivation in Mexico is increasing rapidly. The development in Mexico is limited to half of the low flow of the water which passes through the Imperial Canal; but with the construction of a dam at Black Canyon the flow of the river would be regulated and the extreme low-water periods would no longer exist. This would mean that a large area of Mexican lands, previously neglected, could be brought under cultivation, and Mexican interests undoubtedly would quickly establish rights to a large share of the water which had been stored on American territory at American expense. The Mexican farmers would never have to pay a cent for this great benefit, although with the limited supply of water that exists in the river every increase in Mexican acreage would compel a similar area of American land to remain unirrigated forever. This situation would be especially detrimental to the state of Arizona, where the cultivation of some large areas of land, particularly mesa land, is necessarily postponed to the time when the expected increase in population will justify the expense of cultivation. Much of the Arizona land is at such an altitude that it must be irrigated by pump, while large areas of the Mexican lands may be irrigated by gravity. Therefore Arizona's chances to defeat Mexico in the race to establish water rights based on prior use were considered extremely remote, and agricultural development in Arizona would thus be limited by the small amount of water available above the established Mexican appropriations.

In order to give their argument broader applicability, the opponents stated that Arizona would not be the only part of the United States to suffer from this development. The Mexican lands are held in large tracts, and Oriental laborers are employed at wages far below those of the United States. Taxes are lower there, and water is cheaper.[51] With this combination of advantages, Mexican agricultural products are grown at a very low cost, and are sent to American markets where they can compete successfully with American products. The extension of Mexican acreage would rob the American farmer of more of his markets and would depress prices. Thus the Boulder Canyon Project, which had been meant to promote American welfare, would actually turn out to be an influence detrimental to the prosperity of the country.

The proponents of the Boulder Canyon Project attacked the argument thus stated on several grounds and in addition developed other arguments of their own in an attempt to convince the public that the project would strengthen rather than weaken the

[51] *New Reclamation Era,* Vol. XVII, No. 3 (March 1926), p. 40.

position of the United States as far as relations with Mexico were concerned. They pointed out that the opponents had not mentioned the All-American Canal feature of the project which was designed to solve the Mexican problem and not to complicate the situation. For a number of years the farmers of the Imperial Valley had planned to construct a canal which would lie entirely within American territory. They realized that their continued existence in Imperial Valley depended completely upon a water supply from the Colorado River, and the fact that a large part of the Imperial Canal was located in Mexico placed their future in the hands of a foreign government as long as the Imperial Canal was the sole means of conveying water to the valley. It would be a simple matter to destroy a small section of the canal and to halt the flow of water. Within a few days thereafter the crops in Imperial Valley would be ruined, and the homes and property of the inhabitants would thus be destroyed. The people themselves would have to leave the valley, since there would be no other source of water available even for drinking purposes. It is questionable whether there was any real danger of trouble with Mexico. Nevertheless the fear existed, and the Imperial Valley farmers eagerly supported the project which would bring water to them by an all-American route.

The All-American Canal would not only give the Imperial Valley a more dependable water supply, but it would end the necessity of the agreement with Mexico whereby the Mexican farmers had the right to appropriate half of the water diverted. American lands would have first call on the water flowing through the All-American Canal, and, as far as the American farmers were concerned, the old Imperial Canal could be abandoned. In the past the Imperial Canal had diverted enough water to irrigate about 600,000 acres of land. By 1928 some 400,000 acres were being cultivated in the United States and 200,000 acres in Mexico. But Mexican farmers had continued to expand their acreage. They were able to continue this development, because they were entitled to half of the water. The point had been reached when every increase in acreage cultivated in Mexico meant a corresponding decrease in American cultivation.

The only way that the United States could prevent this shrinkage of cultivated area and secure the fullest development of the Imperial Valley was to build the All-American Canal which would bring an adequate supply of water to the American lands. The new canal would divert water from the Colorado River at a point farther upstream than the heading of the Imperial Canal. It would bring water to the valley at a higher level, thus lowering the pump lifts necessary and increasing the area irrigable by gravity.

The All-American Canal seems to solve the Mexican problem as far as the Imperial Valley is concerned, but it does not stop the appropriation of water for use on Mexican lands. The Imperial Canal is still in existence, and undoubtedly Mexican farmers will continue to irrigate their lands with water diverted from the Colorado River into the Imperial Canal. At a comparatively low cost the Mexican farmers could build a new heading for the canal on Mexican territory. Thus they could continue to divert water and to expand their acreage under cultivation as long as the water was available in the lower river. With an equated flow, a greater amount of water would be available, even though California appropriations through the All-American Canal and the Colorado River Aqueduct were greatly increased. Since California had limited her diversions from the Colorado River, the remainder, which was reserved for Arizona, would flow down the river until such time as Arizona could use it. The question arises, however, whether Mexico has the right to appropriate this water which is reserved for Arizona. The only logical answer to this question is that, once the water has entered Mexican territory, Mexico can use it as she pleases. Mexico cannot prevent the diversion of water by the United States through American territory, nor can the United States prevent the diversion of water by Mexico through Mexican territory. Mexico has sovereignty over her own soil, and, if she wishes, she may divert and use all of the waters of the Colorado River which enter her territory. Still this right does not prevent the future diversion of water by Arizona above the Mexican border. The law of prior appropriation and use does not extend across international boundaries. Even though

Mexico should appropriate the water long before Arizona could use it, Arizona could take the water when she wanted it without regard for the Mexican need. The rules, principles, and precedents of international law impose no duty or obligation upon the United States to deny its inhabitants the use of water lying entirely within its borders, even though that use results in the reduction in the volume of water in the river below the point where it ceases to be entirely within the United States.[52] In 1848 the Treaty of Guadalupe Hidalgo had outlined in Articles VI and VII the respective navigation rights of both countries in the Colorado River, but made no mention of the use of the water for irrigation.[53] Article IV of the Gadsden Purchase Treaty of 1853 superseded these articles, but still reserved no water for irrigation purposes in Mexico.[54] Therefore, with no agreements to limit her, the United States could take, if she wished, all of the water from the Colorado River and leave none for Mexico. Thus the argument of the opponents of the Boulder Canyon Project that the rapid development in Mexico would arrest the development in Arizona does not agree with the accepted principles of international law.

It is generally conceded, however, that the United States will never exercise her right to take all of the water from the Colorado River. Such a move would ruin all of the Mexican agricultural development along the lower river and would probably result in a strained international situation. The Boulder Canyon Project Act specifically stated that nothing in the act is to be construed as a recognition or a denial of Mexican appropriations; but the legislation also provided that any water granted to Mexico must come from the supply available in the lower basin. This second provision indicates that the United States expects to give some water to Mexico, but the amount that Mexico is to receive is still undecided. In the writer's opinion it seems reasonable to conclude that Mexico will be given the right to use enough water to irrigate the present area under cultiva-

[52] 21 Opinions of the Attorney-General 274 (1895). See also *The Schooner Exchange* v. *McFaddon*, 7 Cranch (U.S.) 116, 136 (1812).

[53] *United States Statutes at Large*, Vol. IX, pp. 928–29.

[54] *Ibid.*, Vol. X, p. 1034.

tion, but that she will not be given a right to a quantity of water that would permit her to extend her acreage to the detriment of American lands. The owners of Mexican lands have been accused of opposing the Boulder Canyon Project and of delaying an agreement between the United States and Mexico which would limit Mexico's appropriations of water. These tactics were followed, it is said, to gain time in which to increase the acreage under cultivation. With a larger cultivated area in Mexico, the United States might feel morally bound to grant the Mexican lands a greater quantity of water than they would receive under present conditions with a smaller area irrigated.[55] There is no information available upon which to judge the merits of this accusation. Yet it should be noted that the Boulder Canyon Project is giving flood protection to the Mexican lands which is costing them nothing, and in view of the limited supply of water available, it is very unlikely that the United States will ever feel morally obligated to increase the Mexican appropriation of water above its present status. If Mexico continues to expand her production while the water is still available, she should do so with the knowledge that some day Arizona will probably take her share and that the additional acreage cultivated in Mexico will have to revert to desert.

It would be to the best interests of all concerned and would avoid possible future controversies if the United States and Mexico could reach an agreement at the present time definitely limiting Mexico's rights to Colorado River water. Under similar circumstances an agreement was reached by the United States and Mexico in 1906, when the rights of the two countries to the waters of the Rio Grande were explicitly defined;[56] but so far none of the proposed agreements concerning a division of the waters of the Colorado has been acceptable to both countries. It is unfortunate that an early agreement has not been reached, since a less expensive plan to develop the Imperial Valley might have been followed. It would be much cheaper to enlarge and

[55] House Hearings before the Committee on Rules, 70th Congress, 1st Session, *Boulder Dam* (1928), Statement of Edward T. Taylor, pp. 19–20.

[56] Senate Document No. 357, 61st Congress, 2d Session, *Treaties, Conventions, International Acts, Protocols and Agreements between the United States of America and Other Powers, 1776–1909*, Vol. I, pp. 1202–5.

improve the old Imperial Canal, which follows the natural route to the Imperial Valley, than to build the All-American Canal. If Mexico had agreed definitely to limit her appropriations to the present amount or some other comparatively small figure, the development might have proceeded according to a plan to improve the Imperial Canal and to build a smaller high-line canal along the route of the All-American Canal in order to reach the Coachella Valley and the high lands of the Imperial Valley.[57] As matters stand, the economic justification of the All-American Canal must rest chiefly upon the fact that it enters the valley at a higher level and makes possible the reclamation of more land with fewer pump lifts. With the existing uncertain international situation, however, it was inevitable that the people of the Imperial Valley should demand that a main canal be built entirely within American territory.

Since the Boulder Canyon Project will probably not result in an increase in Mexican cultivation, the argument of the opponents of the project that an increased acreage in Mexico would flood the American market with cheap products and ruin the American farmer is not significant. Mexican competition existed long before the Boulder Canyon Project became a reality, but this competition will probably not increase. It would be poor business on the part of the United States to permit the extension of the cultivated area in Mexico, especially since those lands bear none of the costs of the Boulder Canyon Project and would thus be given an unfair advantage in competition with American products. Yet from a broad economic point of view it is unfortunate that the highly productive cultivation in Mexico must be confined forever to a relatively small area. The irrigated area below the border is an integral part of the Imperial Valley in every sense excepting a political one. The Mexican lands are a part of the same basin. They have the same general location and the same fertility of soil. They have been developed by Americans according to American methods. Their products are sent to American markets, and their supplies are bought on the American side of the border.[58] The greatest nonpolitical

[57] *New Reclamation Era*, Vol. XVII, No. 2 (February 1926), p. 21.

[58] Lewis R. Freeman, *The Colorado River, Yesterday, Today, and Tomorrow* (1923), p. 408.

difference is that the Mexican lands are held in larger tracts, are better financed, and are better farmed. If these same lands were within the United States they would probably be fully cultivated, since they appear to be more productive than large areas of American lands which will receive the benefits of the Boulder Canyon Project. In the normal course of events, without political interference, the most fertile and best-located lands will be cultivated first. But the fate of the Mexican lands was decided long ago when the colonization and development of the North American continent resulted in an international boundary line which happened to divide the lands of the Colorado River delta.

CONSTITUTIONALITY OF THE BOULDER CANYON PROJECT ACT

Legal aspects.—Many of the opponents of the Boulder Canyon Project recognized that federal development and operation of the project would be far superior to state or private control, and a few admitted that the lack of an agreement with Mexico concerning a division of the waters of the Colorado was not a fatal objection. Most of the opponents maintained, however, that the project was an illegal invasion of states' rights and that the Boulder Canyon Project Act was unconstitutional. From the very beginning of the controversy, Arizona had stated that the development would be illegal without her consent, and she had warned the project's proponents that she would carry her fight to the courts if necessary. During the years of discussion that preceded the Boulder Canyon Project Act, therefore, the legal aspects of the project received close attention and every phase of the problem of constitutionality was studied carefully.

One of the most important points of law involved was whether or not the Colorado River is a navigable stream. At first Arizona took the stand that the Colorado was navigable and argued that the states own the stream beds of navigable rivers within their borders. Later Arizona reversed her position concerning the navigability of the river; but her first argument assumed that she owned half of the stream bed of the Colorado where it is the boundary between Arizona and Nevada. Arizona reasoned that a river which is navigable in fact is navigable in law. The Supreme Court of the United States has consistently held that if a stream can be used for commerce or trade in any form, even for the floating of rafts or logs, it is a navigable stream.[59] For many years the Colorado was used to carry commerce, and it is undoubtedly still a navigable river. Since it is navigable, Arizona is the absolute owner of the stream bed where it lies wholly within the state, and to the center of the stream where it constitutes the boundary between Arizona and the states of Nevada and California.[60] There are numerous court decisions to this effect[61] which are based upon the principle that the shores of navigable waters and the soils under them were not granted by the Constitution to the United States but were reserved to the states respectively. The new states have the same rights, sovereignty, and jurisdiction over navigable streams as have the original states. Therefore, Arizona concluded that she controlled one-half of the stream bed where the proposed dam was to be built, and she claimed that the federal government could not build the dam on her property without her consent.

The supporters of the Boulder Canyon Project admitted that the Colorado River is navigable and that the states have jurisdiction over the stream beds of navigable rivers. They pointed out, however, that the state jurisdiction is subject to the rights of the federal government under the "commerce clause" of the Constitution[62] which provides that Congress shall have power "to regulate commerce with foreign nations, and among the several States, and with the Indian tribes."[63] It is generally

[59] *The Genesee Chief*, 12 Howard (U.S.) 443, 457 (1851); *The Daniel Ball*, 10 Wallace (U.S.) 557, 563 (1870); The *Montello*, 20 Wallace (U.S.) 430, 441–42 (1874); *Barney* v. *Keokuk*, 94 U.S. 324, 336 (1876); *Water Power Company* v. *Water Commissioners*, 168 U.S. 349, 359 (1897).

[60] Samuel White, *Memorandum of Law Points and Authorities Respecting the Rights of Arizona in the Colorado River*, prepared and submitted to Hon. George W. P. Hunt, Governor of Arizona (1925), p. 8.

[61] *Martin* v. *Waddell*, 41 U.S. 366, 410 (1842); *Pollard* v. *Hagan*, 3 Howard (U.S.) 212, 229 (1845); *Scott* v. *Lattig*, 227 U.S. 229, 242–43 (1913).

[62] Senate Hearings before the Committee on Irrigation and Reclamation, 70th Congress, 1st Session, *Colorado River Basin* (1928), Statement of Francis C. Wilson, pp. 181–82. See also *Economy Light Co.* v. *United States*, 256 U.S. 113, 221 (1921).

[63] *The Constitution of the United States*, Art. I, Sec. 8, Par. 3.

accepted that under the commerce clause the federal government has the right and power to make improvements upon a navigable river and this authority may be exercised without regard to the consent of the states within the borders of which the river is located. Most streams and water courses are more or less navigable in some of their reaches, and since the nonnavigable sections serve as feeders for and are so closely connected with the navigable sections, it is not practicable to apply a separate rule for each. Where satisfactory regulation of the navigable portions can be secured only through regulation of the nonnavigable sections, the federal government has full regulative power over the entire stream. Therefore, the supporters of the project reasoned that since the river was navigable in its lower reaches the federal government had the authority to build a dam in the Colorado River at Black Canyon or at Boulder Canyon, whether the river is assumed to be navigable or nonnavigable in the canyon section. They reasoned further that on the basis of past decisions, the federal government had the authority to store water and to generate electric energy as an incident thereto in spite of Arizona's objections and her jurisdiction over the bed of the stream.[64]

On the other hand, the opponents argued that the commerce clause should not apply in this case since the purpose of the development was not to improve the navigation of the Colorado River but to secure flood control, the storage of water for agricultural and domestic uses, and the generation of power. They pointed out the fact that there is no provision in the plans for locks or other means of routing commerce around the dam, that actually the dam would be an obstruction to commerce through the canyon, and that the amount of traffic on the river was too small to warrant much attention. They stated that, under the circumstances, the improvement of navigation could not possibly be an important purpose of the project, and they claimed that Congress did not have the right to use its power under the

[64] Rome G. Brown, *The Conservation of Water Powers*, Senate Document No. 14, 63d Congress, 1st Session (1913), pp. 8–9. See also *Green Bay Co.* v. *The Patten Paper Co.*, 172 U.S. 58, 79 (1898); *Green Bay Co.* v. *The Patten Paper Co.*, 173 U.S. 179, 190 (1899).

commerce clause to accomplish other objects not entrusted to the federal government.[65] Apparently, however, the opponents of the project feared that a broad interpretation of the commerce clause would nullify these arguments, and they sought a different basis of reasoning. Thus they reversed their stand concerning the navigability of the river and attacked the legality of the project on other grounds.

The second important point of law involved was whether or not the United States had the right to build the Boulder Canyon Project in violation of the constitutions and laws of the states involved and without their consent. On behalf of the opposition it was claimed that the right to appropriate water for irrigation, power, and other beneficial uses must be obtained from the state and not from the federal government. It was stated that this procedure was recognized by an act of Congress in 1866[66] and was confirmed by later acts, including the Water Power Act of 1920.[67] Thus it would not be within the power of Congress or of six of the Colorado River Basin states to divide and apportion the waters of the Colorado River and its tributaries in which seven states are interested without the consent of the seventh state or a decision of the Supreme Court of the United States. The case of *Kansas* v. *Colorado* (1907) was quoted to the effect that each state has full jurisdiction over lands within its own borders, including the beds of streams and other waters.[68] It was asserted that unless invited to do so the federal government has no authority to interfere with any river, navigable or nonnavigable, in any state excepting to regulate navigation for interstate commerce or perhaps to fulfill international obligations. Therefore it was concluded that every state has the inherent sovereign right to control the uses of water

[65] *McCulloch* v. *Maryland,* 4 Wheaton (U.S.) 316, 423 (1819). See also Brown, *op. cit.,* pp. 8–9.

[66] *United States Statutes at Large,* Vol. XIV, pp. 251–53. See also *Revised Statutes of the United States,* 43d Congress, 1st Session, 1873–1874, Sec. 2339, p. 432, and *The Code of Laws of the United States of America in Force January 3, 1935,* Title 43, Public Lands, Sec. 661, p. 1887.

[67] *United States Statutes at Large,* Vol. XLI, Part 1, pp. 1068–77. See also *ibid.,* Vol. XXXVIII, Part 1, pp. 250–51.

[68] 206 U.S. 46, 93.

and that to deprive a state of this right, especially a state of the arid West, would destroy its autonomy.[69]

This argument was made more specific by Arizona, which claimed that she had inherited the rights that the federal government had possessed in the way of water when Arizona was a territory. Arizona contended that the bed of the Colorado River was not withheld from her by the enabling act, and that the reservation clause of the enabling act does not mention the beds of streams as land to be withheld by the federal government.[70] Arizona insisted that, if Congress had intended to make such a reservation, appropriate language would have been used to make clear this unprecedented intention. The state contended that, in common with certain other states of the arid West, the common law doctrine of riparian rights has never prevailed within her borders.[71] The doctrine of prior appropriation and use has always been the law in that part of the country, and has been recognized as such by Congress.[72] Therefore the United States had no proprietary rights to water even as an owner of riparian lands, and the federal government can acquire the rights to appropriate water for beneficial use only by compliance with the laws of Arizona and the other states affected. On the basis of this reasoning, Arizona concluded that the Boulder Canyon Project would violate her laws concerning the appropriation and ownership of lands and waters within her borders and that, therefore, the project would be illegal and unconstitutional without her approval.

The supporters of the project denied the validity of the opponents' contentions, and claimed that the United States was not completely bound by the water laws of the states. They pointed to the statement of the Attorney General of the United States to the effect that where the legality of diverting and appropriating water for beneficial uses on nonriparian lands is gener-

[69] House Hearings before the Committee on Irrigation and Reclamation, 70th Congress, 1st Session, *Protection and Development of the Lower Colorado River Basin* (1928), Statement of George E. Dern, Governor of Utah, pp. 206–13.

[70] *United States Statutes at Large*, Vol. XXXVI, Part 1, pp. 574–75.

[71] *Constitution of the State of Arizona* (1925), Art. 17, Sec. 1.

[72] *United States Statutes at Large*, Vol. XIX, p. 377. See also *ibid.*, Vol. XXXII, Part 1, p. 390.

ally established, the original right of the government to appropriate surplus water for its own uses, particularly for the reclamation of its enormous holdings of arid lands, has not been surrendered by any act of Congress or divested by the mere creating of states into which those regions have now become incorporated.[73] Most of the territory lying in the Colorado River Basin was acquired by the United States from Mexico under the Treaty of Guadalupe Hidalgo of February 2, 1848,[74] and the federal government became the absolute owner of this territory, subject only to such vested individual rights as might be recognized. This ownership necessarily included both the land and the water.[75] It was argued further that the Constitution of the United States gives Congress the power to dispose of and make all needful rules and regulations respecting the territory or other property belonging to the United States.[76] It was contended that this power is subject to no limitations, and that no state legislation can interfere with this right or embarrass its exercise.[77] Thus it follows that the public waters are a part of the public domain within the meaning of the Constitution and that the federal government may, if it wishes, dispose of the waters apart from the land.[78] The next question to be answered was whether or not the federal government had transferred irrevocably its rights in these waters to the states; and the project's proponents claimed that it had not. They stated that the grant contained in the act of 1866[79] was made primarily for the purpose of protecting the investments made by early settlers in California, who, without authority, had appropriated and used land and water belonging to the United States. They contended

[73] *Annual Report of the Attorney General of the United States for the Year 1914*, p. 39.

[74] *United States Statutes at Large*, Vol. IX, p. 926.

[75] Ottamar Hamele, "Federal Rights in the Colorado River," *The Annals of the American Academy of Political and Social Science*, Vol. CXXXV, No. 224 (January 1928), p. 144.

[76] *Constitution of the United States*, Art. IV, Sec. 3, Par. 2.

[77] *Gibson* v. *Chouteau*, 13 Wallace (U.S.) 92, 99 (1871).

[78] *Cruse* v. *McCauly*, 96 Federal Reporter 369, 373–74 (1899); *Hough* v. *Porter*, 51 Oregon Reports 318, 393 (1909); *Howell* v. *Johnson*, 89 Federal Reporter 556, 558 (1898).

[79] *United States Statutes at Large*, Vol. XIV, p. 251.

that the grant was made to the actual users of water who were complying with local laws and not to the state or other political body. Therefore they concluded that the United States is still the owner of the unappropriated waters and may use them for public purposes without the consent of any of the states.[80]

On behalf of the project it was claimed that as far as Arizona is concerned there is a special reason why that state should not interfere with the project. Arizona entered the Union under an enabling act which expressly retained in the federal government the right to carry on operations of the character of the Boulder Canyon Project. All rights and powers for carrying out the Reclamation Act of 1902 were reserved, and Arizona agreed to disclaim all right and title to the unappropriated and ungranted public lands lying within her boundaries.[81] Congress had the right to reserve water, and if Arizona were still a territory no one would question the right of the United States to deal with the lands and waters as Congress may direct. Since the reservations were made in the enabling act, the power of the federal government to carry out the project in Arizona could not be doubted.

In addition to the refutations and answers to the opponents' contentions, other arguments were developed to prove that the project was legal and that the act was constitutional. It was stated that improvement of navigation was not necessary to bring the act under the commerce clause of the Constitution. This clause does not mention the word "navigation," and the authority under the clause is not limited to navigation alone. Two transcontinental railroads and three transcontinental highways cross the Colorado River below Boulder Canyon. If floods should destroy these bridges, commerce in that part of the Southwest would be paralyzed. Therefore whether the Colorado River is navigable or not, the federal government has the power to regulate the river in order to protect interstate commerce by railroad, truck, etc.

[80] *Camfield* v. *United States,* 167 U.S. 518, 525–26 (1897); *Irving* v. *Marshall,* 20 Howard (U.S.) 558, 561–62 (1857); *United States* v. *Midwest Oil Co.,* 236 U.S. 459, 474 (1915).

[81] *United States Statutes at Large,* Vol. XXXVI, Part 1, pp. 569–70. See also *Constitution of the State of Arizona,* Art. 20, Secs. 4, 10.

It was also argued that the federal government has the power to carry out the Boulder Canyon development in order to protect its own properties. The United States owns the banks of the Colorado River almost exclusively from its source to the Mexican boundary. This property is of enormous value, and it was contended that the federal government has the power to build the dam if for no other reason than to prevent possible floods and other damage to the public domain and to reclaim government lands under the Reclamation Law.[82] The opponents of the project denied the validity of this argument. They stated that the United States government is a government of enumerated powers, and that it does not have the authority to protect and reclaim public lands without the consent of the states.[83] They quoted the decision in the case of *Kansas* v. *Colorado* (1907) in which the court said that if no authority has been granted to the national government, none can be exercised, even though the national government is the only agency in a position to do the work properly.[84] The proponents pointed out, however, that the United States owns the land on both sides of the river at Boulder Canyon and at Black Canyon and that it owns most of the property in the reservoir site. They said that even if the federal government does not have the authority to carry on the project merely to protect and reclaim public lands, federal ownership of the dam and reservoir sites would undoubtedly expedite construction once the authority of the national government had been definitely established on some other grounds.

Another argument which was used to prove the constitutionality of the project was based upon the international character of the river. It was stated that if there were a treaty between the United States and Mexico concerning the regulation of the Colorado River, the federal government would have the authority to do as it saw fit to carry out the treaty obligations.[85] Some of the opponents admitted that the national government had this

[82] Senate Hearings before the Committee on Irrigation and Reclamation, 70th Congress, 1st Session, *Protection and Development of Lower Colorado River Basin* (1928), pp. 303, 306.

[83] Leatherwood, *op. cit.*, pp. 139–40.

[84] 206 U.S. 46, 91–92.

[85] *McCulloch* v. *Maryland*, 4 Wheaton (U.S.) 316, 411–12 (1819).

power, but they pointed out that no treaty with Mexico existed obligating the United States to control and regulate the Colorado River. Still the national obligations existed even in the absence of treaties, and the proponents maintained that the federal government not only had the power to carry out the project for international reasons but that it was the sole political agency legally entrusted with international affairs.[86]

Litigation.—In spite of the opponents' arguments, it seemed likely that the courts would find the Boulder Canyon Project within the scope of federal activities on one basis or another. After the Boulder Canyon Act was passed, however, Arizona made good her threat and started suit to block the project. In 1930 she filed a bill of complaint in the Supreme Court of the United States asking that the Colorado River Compact and the Boulder Canyon Project Act be declared unconstitutional.[87] The complaint filed was directed against the states of California, Nevada, New Mexico, Utah, Colorado, and Wyoming, and against Ray Lyman Wilbur, Secretary of the Interior. It contended that the act was unconstitutional because: (1) it deprived Arizona of her sovereign jurisdiction and control of the water and dam and reservoir sites situated in the state and vested control thereof in the United States without consideration for the laws of Arizona; (2) it subjected Arizona to the Colorado River Compact and made the compact effective in that state without her approval; (3) it authorized the Secretary of the Interior to store the unappropriated water of the river, to withhold the use of this water from Arizona except by contract, and to deprive Arizona of her right to the water by selling it to users outside of the drainage basin; (4) it facilitated the use of the stored water in California and did not extend equal privileges to water users in Arizona; and (5) it authorized the Secretary of the Interior to engage in the business of storing and selling water and generating electric power by the utilization of the natural resources of Arizona without providing appropriate

[86] *Constitution of the United States*, Art. 1, Sec. 10, Par. 1.

[87] Attorney General of the State of Arizona, Motion for Leave to File Bill of Complaint and Bill of Complaint, Supreme Court of the United States, October Term, 1930, *State of Arizona* v. *State of California, et al.*, pp. 3–39.

compensation (such as the privilege to tax improvements) to that state.

Arizona's bill of complaint was accepted by the Supreme Court of the United States on October 13, and the defendants were given until January 5, 1931, to prepare their case.[88] They decided to file motions to dismiss, since such motions, if successful, would lead to an early settlement of the dispute, whereas an answer to the complaint and the resulting trial would probably keep the case in the courts for ten or fifteen years. To support these motions, the defendants filed three arguments, one for Ray Lyman Wilbur, Secretary of the Interior, one for the state of California, and one for the other defendant states. The state of Arizona filed counter-arguments.[89] The grounds assigned in the defendants' motions were: (1) that the bill of complaint did not join the United States, an indispensable party; (2) that it did not present any case or controversy of which the court could take judicial notice; (3) that the proposed action of the defendants would not invade any vested right of the plaintiff or of any of its citizens; and (4) that the bill did not state facts sufficient to constitute a cause of action against any of the defendants. The case was heard on these motions, and the opinion was delivered by Mr. Justice Brandeis on May 18, 1931.

The court held that the mere construction of a dam and reservoir partly located in Arizona would not be an invasion of Arizona's quasi-sovereign rights. This decision was based upon the principle that the United States may perform its functions without conforming to the police regulations of a state. If Congress had the power to authorize the construction of the dam and reservoir, the federal government (represented by Secretary Wilbur) was under no obligation to submit the plans and specifications to the State Engineer of Arizona for approval. In addition, Arizona's argument that the dam, reservoir, and power plant, when completed, would not be subject to the taxing power of the state, could be disregarded. The Boulder Canyon Project Act provided that the title to the works should remain forever in the United States and the exemption from state taxa-

[88] *Engineering News-Record*, Vol. CV, No. 16 (October 16, 1930), p. 628.

[89] *Arizona* v. *California et al.*, 283, U.S. 423, 423–64 (1931).

tion was only an ordinary incident of any public undertaking by the federal government.

It is a well-established point of law that the federal government has the power to build a dam in a river for the purpose of improving navigation if the river is navigable. Arizona's bill of complaint alleged that the river has never been and is not now a navigable river. But the court held that the evidence of history proved that a large part of the Colorado River was formerly navigable and that a river which is navigable in fact is navigable in law. Arizona contended that a motion to dismiss admits every well-pleaded allegation of fact in a complaint and that the defendants had admitted, therefore, that the Colorado was not navigable. The court refused to sustain this contention and held that the court may take judicial notice that a river within its jurisdiction is navigable when the evidence of navigability is so definite as in the case of the Colorado River. Commercial disuse does not amount to an abandonment of the river or prohibit future exertion of federal control. As a matter of fact the navigability of the river would be improved for considerable distances above and below the dam by the creation of the reservoir, the storage of silt, and the regulation of flow.

The court refused to sustain Arizona's complaint that the recital in the Boulder Canyon Project Act of the improvement of navigation as a purpose of the project was a mere subterfuge. The Colorado River Compact had declared navigation of the river to be subservient to other uses; but the court held that it could not inquire into the motives which induced the members of Congress to enact the Boulder Canyon Project Act. It ruled that, since the river is navigable and since the structures provided for in the act were not unrelated to the control of navigation, the construction of the project was clearly within the powers conferred upon Congress. It was not for the court to determine whether or not the particular structures involved were reasonably necessary; and the fact that purposes other than navigation would be served did not cancel the power of Congress to carry out the project, even if those other purposes alone would not have placed the development within the field of federal activity. The court could not assume that Congress did not intend to aid navigation. The act specifically stated that improvement of navigation was one of the primary purposes of the project, and the court could not rule that this statement was governed by the general references to the Colorado River Compact or that the intent of Congress was to abuse its power to regulate navigation. Since Congress had the authority to build the dam and reservoir under its constitutional power to improve navigation, the court held that this case presented no occasion to decide whether the authority to construct the dam and reservoir might not also have been constitutionally conferred for the specified purposes of irrigating public lands, of regulating the flow and preventing the floods in an interstate river, of apportioning the waters among the states equitably entitled thereto, or of performing international obligations. The navigability of the Colorado River, therefore, formed the basis for the constitutionality of the Boulder Canyon Project Act.

Arizona had alleged that the act was not only unconstitutional but that the mere existence of the act would invade her quasi-sovereign rights in respect to the appropriation of waters within or on her borders. The court recognized the fact that Arizona has great need for further appropriations from the river for irrigation, and that the projects planned by the State Engineer would reclaim vast areas of land within the state, including a large area of state-owned land. Nevertheless the court denied Arizona's contention that the act invaded her rights and held that Arizona's argument was based not upon an actual or threatened impairment of her rights but upon an assumed potential invasion. The act provided that it is not to be construed as interfering with perfected rights to water, and that the states may adopt any policies that they deem necessary within their borders with respect to the appropriation of water excepting as modified by interstate agreement. Since Arizona had refused to enter into an interstate agreement, her legal rights were unimpaired by the act and the allocations of water under the Colorado River Compact could not affect her right to appropriate water according to the law of prior appropriation and use.

The court refused to sustain Arizona's allegation that Secretary Wilbur had seized the Colorado River in Arizona and all of the dam sites and reservoir sites suitable for the irrigation of Arizona lands. The court held that there had been no physical taking of possession of anything and that Secretary Wilbur had not trespassed on lands belonging either to Arizona or to any of its citizens. The court pointed out that when the bill of complaint was filed the construction of the dam and reservoir had not been begun and that years would elapse before the project was completed. If Arizona's perfected rights should be interfered with by the operation of the project, when completed, the state could bring her case to court at that time and receive appropriate remedies. In order to promote appropriation, however, Arizona had no constitutional right to use any land of the United States and could not complain of any provision in the act conditioning the use of public lands.

The decision of the court, therefore, upheld the constitutionality of the Boulder Canyon Project Act, and denied Arizona's plea that her sovereign rights were being invaded. The court sustained the motions of the defendants to dismiss Arizona's bill of complaint but dismissed the bill without prejudice to an application for relief by Arizona in case the stored water is used in such a way as to interfere with her perfected rights or with her right to make additional legal appropriations. The court held that this case presented no occasion to consider other questions which had been argued, although one member of the court, Mr. Justice McReynolds, was of the opinion that the motions to dismiss should be overruled and that the defendants should be required to answer.

The decision of the Supreme Court marked the failure of Arizona's last desperate attempt to block the construction of the Boulder Canyon Project. Despite this decision, Arizona for ten years refused to sign the Colorado River Compact and continued to demand that the terms be made more favorable to her interests. She was the only state in the Colorado River Basin still completely free to follow the old rule of prior appropriation and use, and she planned to take every possible advantage of this position.[90] In 1935 Arizona renewed litigation over the

[90] *Acts, Resolutions, and Memorials of the Regular Session, Thirteenth Legislature of the State of Arizona* (1937), Message of Governor R. C. Stanford.

river, and again attacked the apportionment of water among the states. Her petition was denied by the Supreme Court, but the renewal of litigation indicated definitely her hostile attitude. This was an unfortunate situation, since more suits involving interpretations of the Colorado River Compact and the Boulder Canyon Project Act and the respective rights of the states and the federal government in the Colorado River Basin were thus to be anticipated. Arizona's conditional ratification of the compact in 1939 has not, and may not, become effective and her position is still uncertain. Until an effective ratification is secured, the threat of litigation will remain as an obstacle to all future projects which may be planned for the Colorado River Basin.

Chapter IV

CONTRACTS AND FINANCE

One of the most cogent arguments in favor of the Boulder Canyon Project was that the development would pay for itself and that the disbursements made by the national government would be returned to the treasury within a reasonable length of time. The proponents of the legislation had stressed the point that the resulting benefits could be secured without permanent cost to the federal government, and to prove this contention they had devised a financial plan for the development on the basis of a fifty-year amortization period. It is generally conceded that if the project had been proposed as a frozen investment with no possibility of an adequate financial return the Boulder Canyon bill would not have secured favorable action in Congress. The financial features of the project, therefore, occupied an important position in the Boulder Canyon legislation, and, as previously stated, a definite financial plan was incorporated in the act to provide for the repayment of construction, operation, and maintenance costs plus an interest charge for the use of the funds. In general, the financial plan required the Secretary of the Interior to execute contracts for the sale of power and stored water which would insure revenues large enough to meet the amortization payments of the dam and power plant according to the planned schedule. The plan also required the Secretary to enter a contract with the Imperial Irrigation District which would guarantee the repayment of the cost of the All-American Canal and appurtenant works under the provisions of the Reclamation Law. Since the Boulder Canyon Project Act stipulated that these contracts must be executed before any money could be appropriated for construction, the success of the project depended upon whether or not the Secretary of the Interior could secure solvent contracts which would meet the requirements of the financial plan and satisfy the conditions of the act.

COST ESTIMATES

The Secretary of the Interior, Dr. Ray Lyman Wilbur, faced a very difficult task. His problem was to compute rates which would yield an adequate return and to find buyers who were willing to contract for power and water at those rates. Before adequate rates could be computed, however, the costs of construction had to be known. Obviously, a rate schedule to amortize construction costs would be valueless unless the construction costs had been estimated within reasonable limits of accuracy. Since no construction costs had been incurred, Secretary Wilbur was forced to rely entirely upon estimates; and his first move was to verify all cost figures in order to set up a definite goal for revenues.

The Boulder Canyon Project Act provided that the construction costs of the entire project, including the dam, reservoir, powerhouse, machinery, All-American Canal, appurtenant structures, and interest during construction, should not exceed $165,000,000. This figure was the result of exhaustive studies made by the Bureau of Reclamation,[1] supplemented by the review of the Sibert Board, and it represented the best calculation upon which to base the contract computations. Nevertheless this maximum-cost figure could not be used effectively without further analysis; and it was necessary to divide the total cost into its component parts. The costs of the individual structures included in the project were estimated[2] by the Sibert Board as follows:

Dam and reservoir (26,000,000 acre-feet capacity)	$ 70,600,000
1,000,000 horsepower development	38,200,000
All-American Canal	38,500,000
Interest during construction on above	17,700,000
Total	$165,000,000

These estimates were based upon a construction period of seven years. They did not include a branch canal to the Coachella Valley as a part of the project, although the construction of such

[1] Ray Lyman Wilbur and Northcutt Ely, *The Hoover Dam Power and Water Contracts and Related Data* (1933), p. 17.

[2] House Document No. 446, 70th Congress, 2d Session, *Report of the Colorado River Board on the Boulder Dam Project* (1928), p. 7.

a canal appears to have been contemplated by the act. The cost of the branch canal was estimated by the board at $11,000,000, which would increase the total cost to $176,000,000. Still it should be noted that the $17,700,000 estimated interest figure apparently includes an interest charge for the $38,500,000 to be appropriated for the All-American Canal. Since the Boulder Canyon Project Act provided for the construction of the canal under the provisions of the Reclamation Law, no interest charge for canal appropriations should have been computed, and the estimated interest figure should have been considerably lower.

Subsequent investigations, including the careful check of Secretary Wilbur, indicated that the figures given by the Sibert Board were on the conservative side; and, with some minor changes in the plans, the Commissioner of Reclamation, Dr. Elwood Mead, fixed the estimated cost of the dam and power development at $109,446,000, an increase of $646,000 over the $108,800,000 ($70,600,000 + $38,200,000) estimated by the Sibert Board. Interest on $109,446,000 during construction of the dam and power plant was estimated at $11,554,000, thus bringing the assumed cost of this part of the project to $121,000,000. During the negotiations of the contracts, however, the Sibert Board authorized an increase in the height of the dam to raise the water level an additional 25 feet. The increase in cost was estimated at $4,392,000, and the total cost of the dam and power development thus became $125,392,000.[3]

The estimate of $125,392,000 represented the aggregate investment which would be required for the development at Black Canyon, but this figure could not be taken as the net investment to be covered by the revenue contracts. The Boulder Canyon Project Act had provided that $25,000,000 of the cost might be allocated to flood control and repaid out of surplus revenues. In addition, $17,717,000, the estimated cost of the power machinery,[4] might be deducted if the burden of financing the machinery was placed upon the lessees of the power plant. These deductions would reduce the net investment, exclusive of flood control and the cost of machinery, to $82,675,000. On

[3] Wilbur and Ely, *op. cit.*, p. 16.

[4] *Ibid.*

the basis of a 4 per cent rate and an amortization period of fifty years, as required by the act, the interest on this investment would be about $108,107,007. Operation and maintenance expenses and depreciation expenses were estimated at $7,262,557 and $8,875,553, respectively, for the period.[5] The total which the Secretary of the Interior was required to recover through the revenue contracts, therefore, was about $206,920,117, computed as follows:

Repayment (exclusive of flood control and power machinery costs)	$ 82,675,000
Interest charges	108,107,007
Operation and maintenance	7,262,557
Depreciation	8,875,553
Total	$206,920,117

With the acceptance of these estimated costs, the lower limit of the required revenues from the project was established and could be used as a guide in computing the rates for the revenue contracts. But another large item of cost should be mentioned. It was estimated that the transmission lines about 300 miles long which would be necessary to bring the power generated at the dam to the southern California market would cost $50,000,000.[6] This expense would have to be borne by the agencies purchasing the power, and it was not included as a part of the financial plan of the project. Therefore, it was not considered as a cost element to be covered by the revenues derived from the project, but it had to be considered in the computation of rates if competitive conditions were to be met.

CONTRACTS FOR POWER

DETERMINATION OF RATES

The sale of power generated at the dam was expected to account for the greater part of the revenues of the project and to supply most of the funds necessary to cover the costs involved.

[5] House Hearing before the Subcommittee of the Committee on Appropriations, 71st Congress, 2d Session, *Second Deficiency Appropriation Bill for 1930*, p. 1043.

[6] Senate Document No. 186, 70th Congress, 2d Session, *Colorado River Development* (1929), p. 26.

The Boulder Canyon Project Act had offered three alternative methods for the disposal of the power, and a modification of one of these was adopted. The government was to build the powerhouse and install the generating machinery, but the contractors for power were to reimburse the government for the machinery over a ten-year period and were to pay a certain rate for the use of falling water. The power possibilities of the project were so tremendous that there was little doubt of the ability of the project to meet its obligations if a market for the power could be found. Studies of the quantities of water available had indicated that with a dam which would raise the water level 557 feet, 3,600,000,000 kilowatt-hours of firm energy per annum would be available; with an increase in water level to 575 feet, 4,240,000,000 kilowatt-hours; and with an increase to 582 feet, 4,330,000,000 kilowatt-hours.[7] A dam which would raise the water level 557 feet formed the basis of the original calculations of the Sibert Board, but a recommendation for increased height was anticipated. During the negotiation of the contracts, therefore, the 4,240,000,000 figure was assumed; but when the board made its decision the water-level height was increased to 582 feet, thus creating an additional firm power output of 90,000,000 kilowatt-hours, which was disposed of separately. With this height of dam it was estimated that some 1,550,000,000 kilowatt-hours of secondary energy would also be generated, and the disposal of this power formed another possible source of revenue to be considered in the drafting of the power contracts.

Under the provisions of the Boulder Canyon Project Act, the rates charged for power were to be high enough to amortize the cost of the dam and power development but were not to be above the rates determined by competitive conditions at distributing points or competitive centers. The southern California market was the only available market of sufficient size to absorb enough of the Boulder Canyon power to yield the required revenues. That market was located over large oil and gas deposits, and the cost of steam-generated energy provided by the use of oil and gas as fuel would, therefore, probably fix the competitive value of the power. In order to compute this competi-

[7] Wilbur and Ely, *op. cit.*, p. 16.

tive value several studies were undertaken to estimate the cost of producing and marketing a similar quantity of steam power under existing conditions in southern California, and these studies led eventually to the determination of a rate for Boulder Canyon power which could compete with other methods of power production. According to the best estimates available this rate would also produce enough revenue under the contracts to meet the amortization provisions of the act and would thus fulfill the rate requirements from both the competitive and the revenue points of view.

The power cost estimates were completed before the Sibert Board authorized an increase in height of the dam; and it was assumed that $121,000,000 would be the cost of the Boulder Canyon development. From this figure was subtracted the $25,000,000 allocated to flood control; but to it was added $50,000,000 for the cost of transmission lines. Thus there was a net addition of $25,000,000 which brought the total estimated cost of the power development to $146,000,000, or to about $195 per kilowatt on the assumption of an installed capacity of 750,000 kilowatts. According to the various studies none of the other proposed hydroelectric power developments could supply the southern California market at as low a cost per kilowatt of installation. Competitive hydroelectric power was ruled out as a consideration, and attention was focused upon steam power as a limiting factor to the value of Boulder Canyon power. Therefore in order to estimate the proper rate it was necessary to forecast future fuel-oil costs, steam-power-plant construction costs, transmission costs, and steam stand-by costs, and to give consideration to the competitive value of Boulder power at various load factors under assumed production rates and assumed transmission losses.

The price of fuel oil (or of natural gas) is the most important single factor affecting the cost of steam power in southern California. In 1929 the price of natural gas was so low that its cost as boiler fuel was equivalent to oil at about 50 cents per barrel.[8] Owing to the depletion of near-by fields, however,

[8] On May 28, 1929, the Governor of California had approved a law which forbade the waste of natural gas taken out of the ground in the process of developing

such low costs were not expected to continue, and the power-cost estimates provided for an increase. The Boulder Canyon Project Act stated that the rates for the sale of power might be readjusted fifteen years after the contracts were executed and every ten years thereafter. Therefore the fuel-oil price considered in determining the competitive value of Boulder Canyon power for the original contract purposes was the average price for the period beginning with the production of the power and ending fifteen years after the execution of the contracts. This average price was estimated to range from 75 to 80 cents per barrel.[9] Since a difference of 5 cents per barrel in the assumed price of oil would result in a difference of about $325,000 per year in the cost of steam-generated power at a load factor of 60 per cent, the accuracy of the price estimates was an important consideration in determining the value of Boulder Canyon power; but in order to place the cost forecasts within conservative limits the 75-cent figure was used in the computations.

The power studies were based on an annual production of 3,600,000,000 kilowatt-hours, and thus with an estimated transmission loss of 12 per cent the amount of power to reach Pacific Coast terminal substations was taken as about 3,168,000,000 kilowatt-hours. Actually a larger amount would be generated with the increased height of dam, and the sale of this additional product would provide a substantial margin of safety in the financial setup.[10] An annual income of $375,000 from the sale of stored water was included in the computations which lowered slightly the selling price for power necessary to yield the required revenues from the project;[11] but no adjustment was made for the sale of secondary energy, although the contracts

oil. Under this law markets for a large quantity of natural gas had to be developed immediately, and since oil operators did not wish to shut down profitable wells the price of natural gas became very low. Electric-power companies in the southern part of the state within easy transmission distance of the oil fields took advantage of this low-cost fuel, and the cost of production of steam-generated power became unusually low. See *Statutes of California* (1929), chap. 535, p. 927. See also *Engineering News-Record,* Vol. CIII, No. 17 (October 24, 1929), p. 638.

[9] Wilbur and Ely, *op. cit.,* pp. 490–91.

[10] California Colorado River Commission, *Colorado River and the Boulder Canyon Project* (1931), Table C-1, p. 164.

[11] *Ibid.,* Table 1-x, p. 172.

provided that this power should be sold at the rate of 0.5 mill per kilowatt-hour.[12] Revenues from the sale of secondary energy, therefore, would provide still another element of safety in the financial plan.

Under these assumptions, with an installation of 12 units having a capacity of 750,000 kilowatts, the cost of Boulder Canyon power at the dam was estimated at 1.627 mills per kilowatt-hour; and 1.63 mills was taken as the contract price. At this price the cost of the power at the Pacific Coast terminal substations was computed as 3.921 mills per kilowatt-hour, which was slightly higher than the cost of 3.920 mills per kilowatt-hour estimated for steam-generated power at the same terminal substations. Actually, however, the price for steam power would probably not be as low as the estimate, since that figure was based upon a conservative price for fuel oil and upon the assumption that all steam power would be generated in modern, efficient, and well-located plants. The price of 1.63 mills would meet, therefore, the requirement of the Boulder Canyon Project Act that the rates charged must not be above the rates established by competitive conditions at distributing points or competitive centers.

Additional revenue from power was expected to be derived from the sale of secondary energy, which was defined as all power generated in excess of the amount defined as firm energy. This power would probably not be available continuously or in accurately predictable quantities, since its production would depend upon whether or not enough water was available in the reservoir to generate power in excess of firm energy. Such power, therefore, was not in an advantageous position to command a high price, and it would have to sell for any price which it would bring under competitive conditions. Still, if the price were high enough to cover the separable costs of production of the secondary energy and to contribute something to the joint costs, it would be worth while to generate this power in excess of firm energy. During the negotiations with the prospective power purchasers, the price of 0.5 mill per kilowatt-hour was determined as a fair price for secondary power on the basis of

[12] Wilbur and Ely, *op. cit.*, pp. 141–42.

TABLE VII.—ESTIMATED COST OF BOULDER CANYON POWER*

Assumptions:[a]		
Number of units installed in power plant		12
Installed capacity in kilowatts		750,000
Load factor in percentage		55
Output in millions of kilowatt-hours per year		3,600
Revenue per year from domestic water		$ 375,000
Cost:		
Dam, intake works, tunnels, etc.		$ 86,720,907
Power-plant building and inclined railway		2,679,162
Interest during construction on above items		11,048,899
Total cost including interest during construction		$100,448,968
Flood control		25,000,000
Total cost less allocation for flood control		$ 75,448,968[b]
Annual charges:		
Operation and maintenance of dam		$ 133,993
Depreciation of dam		156,940
Annuity to cover interest and repayment of all excepting $25,000,000 for flood control		3,512,165
Subtotal (annual charges without surplus)		$ 3,803,098
Unit cost (mills) per kilowatt-hour	1.056	
Unit cost (mills) per kilowatt-hour, plus 10% for contingencies	1.162	
Annuity to cover interest and repayment of $25,000,000 flood control		1,163,755
Surplus to Arizona and Nevada (3/5 of above)		698,253
Total annual charges		$ 5,665,106
Unit cost (mills) per kilowatt-hour	1.574	
Unit cost (mills) per kilowatt-hour, plus 10% for contingencies		1.731
Reduction in cost (mills) of power due to annual revenue of $375,000 from domestic water		.104
Net cost (mills) of Boulder Canyon power		1.627

* California Colorado River Commission, *Colorado River and the Boulder Canyon Project* (1931), Table C-20, p. 171.

[a] Dam and Powerhouse constructed by the government, and power-plant machinery and equipment purchased, installed, and operated by lessees.

[b] This figure does not agree with the net investment figure of $82,675,000 given on p. 140, above. There are three reasons for the discrepancy: (1) the $82,675,000 includes an allowance for $4,392,000 for the increased height of the dam which was not included in the McClellan Report; (2) the cost of power-plant machinery and equipment to be borne by the lessees is estimated by the McClellan Report as $20,295,971 instead of $17,717,000 (see California Colorado River Commission, *op. cit.*, Table C-16, p. 170); and (3) these changes naturally caused differences in estimated interest, operation and maintenance, and depreciation charges.

TABLE VIII.—ANNUAL COST OF BOULDER CANYON POWER AT PACIFIC COAST TERMINAL SUBSTATIONS COMPARED WITH STEAM-GENERATED POWER

Boulder Canyon Power	
Cost of falling water (3,600,000,000 kilowatt-hours @ 1.63 mills per kilowatt-hour)	$ 5,868,000
Generating plant costs[a]	1,943,308
Transmission costs[b]	3,917,000
Steam stand-by costs (oil, $0.75; 55 per cent load factor)[c]	521,000
Transmission-line cost from steam stand-by[d]	171,000
Total annual cost at terminal substation	$12,420,308
Cost (mills) per kilowatt-hour (3,168,000,000 kilowatt-hours) Boulder Canyon power	3.921
Steam-Generated Power	
Generating plant costs (oil, $0.75; 55 per cent load factor)[e]	$11,397,000
Transmission costs[f]	1,020,000
Total annual cost at terminal substation	$12,417,000
Cost (mills) per kilowatt-hour (3,168,000,000 kilowatt-hours) steam-generated power	3.920

[a] California Colorado River Commission, *op. cit.*, Table C-16, p. 170.
[b] *Ibid.*, Table C-5, p. 165. [c] *Ibid.*, Table C-9, p. 167. [d] *Ibid.*, Table C-11, p. 167.
[e] *Ibid.*, Table C-2, p. 164. [f] *Ibid.*, Table C-7, p. 166.

cost of production and of fair competitive price at the load center. The rate for secondary energy was considered at the same time as the rate of 1.63 mills for firm energy, and it was determined in the same manner.[13] The next step was to secure contracts which would supply large enough revenues at these rates to fulfill the provisions of the financial plan.

DRAFTING OF THE POWER CONTRACTS

On September 10, 1929, the Department of the Interior sent notices to all prospective purchasers of Boulder Canyon power that their applications for such power must be filed not later than October 1. Twenty-seven applications were received, but some of these were conditional and others were indefinite.[14]

[13] Letter from the United States Department of the Interior, Bureau of Reclamation, May 9, 1938.

[14] Wilbur and Ely, *op. cit.*, pp. 18, 515.

There were four principal applicants. The City of Los Angeles and the Southern California Edison Company each asked for the entire power output, which was assumed at that date, prior to the final decision on the height of the dam, to be 3,600,000,000 kilowatt-hours. The Metropolitan Water District[15] asked for about half that amount of energy, and the state of Nevada asked for a third of it. The total of all applications was over three times the quantity of power available; and Secretary Wilbur was thus faced with the problem of allocating the energy among the conflicting applicants.

The allocation was undertaken on the premise that the first requisite was to protect the public interest by providing adequate security for the taxpayers' money. It was recognized that a broad regional distribution of the power would be desirable but that neither Arizona nor Nevada was in a position to make immediate use of the power within her borders. The California applicants, on the other hand, could send the power through channels where it was already being distributed, and would avoid in this way any sharp business disturbance. If all of the Boulder Canyon power were allocated to California, however, the problem of power allocation among the California applicants still remained to be solved. In reaching his decision the Secretary of the Interior decided that the public interest would be served best by giving consideration to the urgent water needs of the Los Angeles area. The first demands to be recognized, therefore, were those of the Metropolitan Water District for sufficient power to pump water from the Colorado River through an aqueduct to the Pacific Coast.[16]

On October 21, 1929, Secretary Wilbur announced a tentative plan for the allocation of Boulder Canyon power. Under

[15] The Metropolitan Water District of Southern California was formed for the purpose of bringing water from the Colorado River to Los Angeles and the cities adjacent thereto under the Metropolitan Water District Act of California approved May 10, 1927, and amended June 10, 1929. The district was comprised of the cities of Los Angeles, Santa Monica, Beverly Hills, Pasadena, Santa Ana, Anaheim, San Marino, San Bernardino, Colton, Glendale, and Burbank. See California Colorado River Commission, *Colorado River and the Boulder Canyon Project* (1931), p. 207. See also *Statutes of California* (1927), chap. 429, p. 694, and *Statutes of California* (1929), chap. 796, p. 1613.

[16] Wilbur and Ely, *op. cit.*, p. 18.

this plan the Metropolitan Water District was allocated 50 per cent of the power, or so much thereof as might be needed for the pumping of Colorado River water to the Los Angeles area. The City of Los Angeles was allocated 25 per cent, and the Southern California Edison and associated companies were allocated the remaining 25 per cent.[17] These allocations were subject to certain deductions and reservations, including a maximum of 18 per cent each for Arizona and Nevada and 4 per cent for the municipalities which had filed applications. This provision for possible deductions was made necessary by the allowance for the preference rights listed in the Boulder Canyon Project Act. All such preference rights were required to be exercised through the execution of valid contracts, and rules were drawn for the reallocation of unused power available if the states, the municipalities, or the Metropolitan Water District failed to absorb their quotas. If all of the withdrawals were made, the original contractors would suffer a considerable reduction in the amount of power available for them; but the exercise of only a small percentage of the preference rights was expected. Provision was made for the creation of a board of control whose members would be appointed by the City of Los Angeles, the Metropolitan Water District, the Southern California Edison Company, and the Secretary of the Interior to act with the City of Los Angeles in the operation of the power plant. The federal government promised to build the dam, tunnels, powerhouse, and penstocks; but the machinery for the generation and distribution of power was to be provided and installed by the lessees. The costs of installation and operation were to be borne by those contracting for the power in proportion to the amounts received; but after operation of the powerhouse had begun the rates charged might be changed either upward or downward in accordance with the actual costs and the provisions of the Boulder Canyon Project Act. The charge for storing water for the Metropolitan Water District was placed at 25 cents per acre-foot.[18]

On November 12 and 13, 1929, hearings were held in order to give the conflicting applicants an opportunity to protest against

[17] *Engineering News-Record,* Vol. CIII, No. 17 (October 24, 1929), p. 670.

[18] *New Reclamation Era,* Vol. XX, No. 11 (November 1929), p. 170.

the proposed allocation, to ask questions, and to make statements. The municipalities protested that 4 per cent of the power was not enough for them. The City of Los Angeles pointed out that after the states and the Metropolitan Water District had been provided for, there would be only a small share of the power left, and demanded that the city applicants should receive the total balance on the basis of their preference rights as opposed to the applications of the private power companies. The private power interests contended, on the other hand, that the development should aim toward the most effective utilization of the resources and that they were in a better position than the cities to distribute the power in the country districts, where the need was greater than within the municipalities themselves. The state of Nevada protested that it should receive a greater allotment for future development, and claimed that its preference right entitled it to power even though the energy could not be used immediately within the boundaries of the state. All of the municipalities and the district urged that the board of control to operate the power plant should be representative in its membership in proportion to the power allocations and should be merely advisory in capacity. They argued that a single operating agency responsible to the Secretary of the Interior would provide more efficient power production and that a divided authority would greatly reduce the market value of the power because of the increased steam stand-by which would be required at the place of use. The task of deciding whether or not these protests were based on valid objections and whether or or not a new allocation schedule would be necessary was thus added to the problems confronting the Secretary of the Interior.[19]

Secretary Wilbur announced that he did not intend to complete the contracts at that time, since he wished to give Arizona and California one more chance to reconcile their differences in order to avoid the discord that had marked most of the previous proceedings in connection with the Boulder Canyon Project. The representatives of the states met in January 1930 but

[19] E. F. Scattergood, "The Status of Boulder Canyon Power Allocations," *The Annals of the American Academy of Political and Social Science*, Vol. CXLVIII, No. 237, Part 2 (March 1930), pp. 39–40.

again failed to find a plan acceptable to all parties concerned. Active negotiation of the power contracts was resumed during the latter part of February 1930, and an agreement concerning the power allocations was reached on March 20.[20] The new agreement was stated in terms of minimum use, with provisions for additions, as opposed to the tentative allocations of October 21, 1929, which were stated in terms of maximum use subject to certain deductions and reservations. The new agreement proposed that 36 per cent of the first power from Boulder Canyon be allocated to the Metropolitan Water District, 13 per cent to the City of Los Angeles, 6 per cent to the other municipalities which had filed applications, and 9 per cent to the Southern California Edison Company. If the municipalities failed to use their entire share of the power allocated to them, the excess was to be given to the City of Los Angeles. In this way 64 per cent of the power was divided among the most important applicants with the belief that it could be placed into immediate use when generated. The allocation of the remaining 36 per cent, or the unused firm power, was more complicated. The Metropolitan Water District was to have first call on this power and all secondary power up to its total requirements for pumping water into and through the aqueduct; and the remainder, if any, was to be divided equally between the City of Los Angeles and the private power interests. Subject to the first call of the Metropolitan Water District, the city was to have the right to 50 per cent of the secondary energy generated, the Southern California Edison Company to 40 per cent, and the smaller utility companies to 10 per cent.

No mention had been made in the agreement of an allocation of power to Arizona and Nevada, and it was necessary to make a provision for them. According to the final allocation each of these states received the right to use 18 per cent of the total firm energy, and thus the 36 per cent unused firm energy might be absorbed by the states if they elected to exercise their preference rights. Either of the two states was given the right to absorb 4 of the 18 per cent allotted to the other state if it were

[20] Wilbur and Ely, *op. cit.*, p. 20.

not in use within a certain period. The Southern California Edison Company agreed to yield energy to three associated companies on the basis of a future agreement among the four companies.[21] Ultimately two of the companies, the Los Angeles Gas and Electric Company and the Southern Sierras Power Company contracted, and each was allocated 0.9 per cent of the firm power, leaving a minimum of 7.2 per cent for the Southern California Edison Company. In addition they were given rights to certain shares of unused power as shown in Table IX.

The contracts with the City of Los Angeles, the Metropolitan Water District, and the Southern California Edison Company were signed on April 26, 1930. These contracts were made operative for fifty, forty-nine, and forty-seven years, respectively; but allowances were made for possible readjustments of rates to accord with competitive conditions under the provisions of the Boulder Canyon Project Act. The decrease in the amount of water available for the generation of power caused by upstream developments and the gradual silting of the reservoir was also considered, and the corresponding shrinkage in kilowatt-hours produced was estimated at 8,760,000 per annum.[22] The contracts bind the city, district, and company to take 100 per cent of the firm power generated if it is available; but they are subject to certain privileges of the municipalities and states, as explained above.

On April 27, 1930, the 11 smaller municipalities reached an agreement to divide the 6 per cent of firm energy allocated to them in proportion to their consumption of energy in 1929.[23] Ultimately (September and November, 1931), however, only Burbank, Pasadena, and Glendale elected to contract, and the remainder of the municipalities' allocation was absorbed by the City of Los Angeles. On November 5 and 12, 1931, the Southern Sierras Power Company and the Los Angeles Gas and Electric Corporation contracted for power and received their share from

[21] Wilbur and Ely, *op. cit.*, pp. 21–23.

[22] *Ibid.*, p. 17.

[23] The 11 municipalities were Burbank, San Bernardino, Pasadena, Glendale, Riverside, Santa Ana, Newport, Beverly Hills, Colton, Anaheim, and Fullerton. See Wilbur and Ely, *op. cit.*, pp. 531–32.

TABLE IX.—DISPOSITION OF BOULDER CANYON POWER*

	FIRM ENERGY			SECONDARY ENERGY
	Minimum Which United States Must Supply (*Percentage*)	Contractor's Obligations if Energy Is Available (*Percentage*)	Maximum Which Contractor Can Demand under Various Conditions (*Percentage*)	
Arizona	18.0000		22.0000	None
Nevada	18.0000		22.0000 (18% plus 4% not used by other state)	None
Metropolitan Water District	36.0000	36.0000	72.0000 (its own minimum, plus first call on unused state energy)	First call on all secondary energy
Los Angeles	14.9054 (13% plus uncontracted municipal energy)	32.9054 (its minimum plus ½ unused state energy, subject to Metropolitan's first call)	32.9054	Call on ½ secondary energy subject to Metropolitan's first call

Pasadena	1.6183	1.6183	1.6183	None
Glendale	1.8867	1.8867	1.8867	None
Burbank	0.5896	0.5896	0.5896	None
Southern California Edison Company	7.2000	21.6000 (7.2% plus 40% unused state energy)	21.6000	Call on 40% of secondary energy, subject to Metropolitan's first call
Los Angeles Gas and Electric Corporation.	0.9000	2.7000 (0.9% plus 5% of unused state energy)	2.7000	Call on 5% of secondary energy, subject to Metropolitan's first call
Southern Sierras Power Company	0.9000	2.7000 (0.9% plus 5% of unused state energy)	2.7000	Call on 5% of secondary energy, subject to Metropolitan's first call
Total	100.0000	100.0000	......	....

* Ray Lyman Wilbur and Northcutt Ely, *The Hoover Dam Power and Water Contracts and Related Data* (1933), p. 25.

the Southern California Edison Company's allocation. Thus of the 27 original applicants only eight actually agreed to take Boulder Canyon power, and on these eight contractors rests the chief financial burden of the Boulder Canyon Project. The contracts were so arranged that the performance of any two of the three principal contractors would yield revenues sufficient for amortization requirements.[24] Since the contracts will account for over $327,000,000 in revenues during the fifty-year period, they represent one of the largest power transactions in the history of the world.

LEGAL ASPECTS AND CONTROVERSIAL POINTS

The contracts of April 26, 1930, obligating the three principal contractors to pay for all of the firm power generated at the Boulder plant were drawn as two rather than three separate documents. The first was a contract for the lease of power privileges by the United States to the City of Los Angeles (through its Department of Water and Power) and the Southern California Edison Company as several, not joint, lessees. The second was a contract between the United States and the Metropolitan Water District for the purchase of electric energy. These contracts required the city and the company to generate energy at cost for the Metropolitan Water District and the other five allottees. Neither Arizona nor Nevada had taken advantage of her right to 18 per cent of the total firm power, but the way was kept clear for the states to exercise their option at any time within fifty years.[25]

The state of Arizona opposed the Boulder Canyon revenue contracts as bitterly as she had opposed the Boulder Canyon Project Act. At the beginning of the controversy Arizona had demanded the right to tax that part of the project which would be constructed within her borders and had objected to the provision which would give to her a revenue from the project in lieu of taxes. Obviously, the amount which Arizona would receive would depend largely upon the rates charged for power and

[24] Wilbur and Ely, *op. cit.*, pp. 23–24. For copies of the power contracts and an analysis of their contents, see *ibid.*, pp. 43–68, 115–288.

[25] *Ibid.*, p. 45.

TABLE X.—FIRM-POWER REVENUES OF THE BOULDER CANYON PROJECT*

Assumptions:

Revenue from 100 per cent of firm energy only
No revenue from sale of water
No revenue from sale of secondary energy
Machinery investment repaid separately by lessees of power plant within ten years
Repayment period, fifty years

Gross revenue from sale of firm energy at 1.63 mills per kilowatt-hour:		
City of Los Angeles	$121,310,549	
Metropolitan Water District	118,031,886	
Southern California Edison Company	88,523,915	$327,866,350
Distribution of revenue for 50-year period:		
Operation and maintenance	$ 7,262,557	
Depreciation	8,875,553	
Interest charges on all except the $25,000,000 allocated to flood control	108,107,007	
Repayment (exclusive of flood control)	82,674,907	
Interest charges on accumulated deficit	63,973	206,983,997
		$120,882,353
Surplus:		
18¾ per cent to Arizona	$ 22,665,441	
18¾ per cent to Nevada	22,665,441	
Interest charges on flood control	20,981,303	
Repayment of flood control	25,000,000	91,312,185
Surplus (available for general development on the Colorado River)		$ 29,570,168

* House Hearing before the Subcommittee of the Committee on Appropriations, 71st Congress, 2d Session, *Second Deficiency Appropriation Bill for 1930*, Statement of Secretary Wilbur, p. 950.

water, and the state demanded a more definite figure—although it was pointed out that if a minimum revenue were granted, a maximum should also be set, corresponding approximately with the amount which would be derived under the application of the highest tax rate in force in the state.[26] However, it was a well-established point of law that such property could not be taxed

[26] Ralph L. Griswell, "Colorado River Conferences and Their Implications," *The Annals of the American Academy of Political and Social Science*, Vol. CXLVIII, No. 237 (March 1930), Part 2, p. 17.

without the consent of the federal government, and since Congress had the power to provide for payments in lieu of taxes,[27] Arizona had no legal grounds upon which to object to this phase of the planned procedure.

Another objection was raised concerning the right of the national government to impose a charge for the use of the stored water and the electric power developed on the project. The Senate Committee on the Judiciary, when directed to study this question, came to the conclusion that Congress had the proper authority to impose the charge and that the general plan to lease or sell the water and energy privileges would not be beyond the constitutional powers of the federal government.[28] When the plan to finance the project through contracts for water and power was incorporated in the Boulder Canyon Project Act, therefore, the proponents of the project were firmly convinced that the financial provisions rested upon a sound legal basis.

During the period devoted to the determination of rates, the allocation of power, and the drafting of the contracts, the Secretary of the Interior sought the advice of the Attorney General of the United States in order to be sure that the financial provisions of the Boulder Canyon Project Act were being interpreted correctly and fulfilled properly. Questions had arisen concerning interest charges for the advances made to build the All-American Canal, the amortization of the $25,000,000 allocated to flood control, and the interest charges on the $25,000,000. On December 26, 1929, the Attorney General rendered the opinion that the advances from the general treasury to the Colorado River Dam Fund for construction costs of the All-American Canal were not interest-bearing, since that part of the project had been placed definitely under the provisions of the Reclamation Law. He decided also that the Secretary of the Interior was not required, in fixing rates, to make provision for the amortization within fifty years of the $25,000,000 allocated to flood control, but that interest should be charged. It was clearly

[27] Congress has provided for payments to the states in lieu of taxes in other instances. See, for example, the *United States Statutes at Large*, Vol. XXXV, Part 1, p. 260; Vol. XXXVII, Part 1, p. 843; Vol. XLI, Part 1, pp. 450, 1072–73.

[28] Senate Hearings before the Committee on Irrigation and Reclamation, 70th Congress, 1st Session, *Colorado River Basin* (1928), p. 475.

the intent of Congress that the principal sum should be repaid out of 62½ per cent of the excess revenues, if any, and it followed that interest charged on the flood control allocation should also be paid from 62½ per cent of the excess revenues and should not constitute an absolute charge to be considered in the determination of rates.[29] Further advice concerning the interpretation of the language of the Boulder Canyon Project Act was sought from the Solicitor of the Department of the Interior, who consolidated his opinions concerning the act in a report dated January 6, 1930.[30] The Secretary of the Interior carefully followed the recommendations of both the Attorney General and the Solicitor, and eventually the financial plan was developed to the point where it seemed to fulfill the requirements of the act in the opinion of all interested parties except the state of Arizona.

The Boulder Canyon revenue contracts, as finally drafted, reserved 18 per cent of the firm energy for Arizona to be taken any time within fifty years and also provided revenues which would yield the state, according to estimates, from $22,000,000 to $31,000,000 during the life of the contracts.[31] In spite of these benefits, Arizona continued to oppose the development and tried unsuccessfully to block the first appropriation to begin construction on the project. She accused California of attempting to rush through the completion of the contracts with the federal government in order to eliminate the consideration of both power and water revenues from any future negotiations that the two states might enter into in attempting to reach an agreement concerning the Colorado River Compact. This accusation was very unreasonable, since Arizona had been given every opportunity to express an opinion and to make suggestions concerning the contracts but had elected to employ obstructionist tactics instead. Secretary Wilbur in his letter of May 9, 1930, to Governor John G. Phillips of Arizona, pointed out that the complaint of haste could not be taken seriously.[32] The federal government had waited some sixteen months, from January 1929 until April

[29] 36 Opinions of the Attorney General 121, 127–45.

[30] Wilbur and Ely, *op. cit.*, pp. 629–42.

[31] *Ibid.*, p. 26.

[32] *Ibid.*, p. 605.

1930, before completing the contracts in order to give the states ample time in which to work out their problems. Arizona had been invited three times to attend the contract conferences but had failed on each occasion to send a representative or even to apply for a share of the power. An agreement between Arizona and California seemed hopeless, and the Secretary felt that for the good of the entire Southwest the project should be delayed no longer. In spite of Arizona's attitude the contracts were drawn to provide both power and revenue for her benefit. It was quite clear that Arizona had been given more consideration than she should have expected under the circumstances, and in the Secretary's opinion the federal government was more than justified in proceeding with the contract and construction plans without Arizona's approval.

After the contracts had been signed Arizona based her objections upon the claim that the lease contract with the City of Los Angeles and the energy contract with the Metropolitan Water District were not valid contracts. The Boulder Canyon Project Act required that before an appropriation could be made for the project the Secretary of the Interior must have secured contracts enforceable at law which would in his opinion provide revenues adequate to reimburse the United States. Arizona admitted that the adequacy of the revenues was left to the judgment of the Secretary, but insisted that two of the contracts were void for other reasons. Arizona claimed that the City of Los Angeles did not have the legal capacity to assume the obligations which the lease contract would impose upon it unless two-thirds of the voters approved of the action as provided by the Constitution of California,[33] and that since the approval had not been secured the contract was unenforceable at law. Arizona claimed further that the lease did not impose any absolute obligation upon either the city or the company to take or pay for energy, and that the obligations imposed were merely conditions of preserving the allocations—since the contract lacked mutuality of obligation, it could not be a valid contract; and if it were not a valid, enforceable contract, it could not fulfill the requirements of the Boulder Canyon Project Act.

[33] *Constitution of the State of California,* Art. 11, Sec. 18.

Arizona also attacked the legality of the energy contract with the Metropolitan Water District on the ground that it was void for want of mutuality and was not a true contract. Arizona stated that the district was not obligated to take or pay for any electric energy, since the district was required to absorb only as much power as might be needed to pump Colorado River water through the aqueduct to the Pacific Coast. At the time the contract was signed the aqueduct had not been built and even a pumping station had not been constructed. The district might never take any of the power, and, therefore, it was maintained, the agreement lacked consideration on the part of the district. Since a bilateral contract must have consideration, or mutuality of obligation, to be a valid contract, Arizona concluded that the Metropolitan Water District energy contract must be void.[34]

Testimony refuting Arizona's position was presented on behalf of the contractors and the Department of the Interior. It was pointed out that the City of Los Angeles had the authority to make the necessary contracts under the provisions of its charter, adopted January 22, 1925, and amended January 18, 1927, and January 15, 1929; and that the Metropolitan Water District had the authority under the Metropolitan Water District Act of the California Legislature approved May 10, 1927, and amended June 13, 1929.[35] The proponents of the project decided, however, in view of the brief time remaining before the adjournment of Congress and the possibility of a filibuster, to eliminate the Arizona objections by amendment of the contracts. The amendment to the lease contract was signed on May 28, 1930, and was supplemented by a second amendment on September 23, 1931. The department of water and power of the City of Los Angeles was given the authority to act in behalf of the city and also as a principal in its own behalf for contract purposes. This amendment sought to avoid the objec-

[34] House Hearing before the Subcommittee of the House Committee on Appropriations, 71st Congress, 2d Session, *Second Deficiency Appropriation Bill for 1930*, pp. 1168–75, 1178–86.

[35] *Ibid.*, pp. 1009, 1036, 1158. See also *Statutes of California* (1927), chap. 429, p. 700.

tion that the city lacked the capacity to contract without approval of two-thirds of the voters. The allocations of energy to the district, city, and company were also amended to state more specifically the amounts the contractors would be required to take, and the minimum annual payments demanded in order to provide definite consideration for the contracts.[36] On May 31, 1930, the allocation provision of the contract with the Metropolitan Water District was amended to require the district to take at least a certain minimum amount of energy and thus to supply the mutuality of obligation said to be lacking by the opponents of the project.[37] The amendments effected no change in the tenor of the contracts and were made chiefly to avoid unnecessary litigation.

Arizona claimed that the amendments had not improved the situation and contended that the contracts were still void and unenforceable. The contracts were submitted to the Attorney General for an opinion, and on June 9, 1930, he reported that the contracts fulfilled the requirements of the Boulder Canyon Project Act which were made conditions precedent to the appropriation of money for construction.[38] He pointed out that the contracts did not violate the California Constitution, since the approval of two-thirds of the voters was required only when a city assumed liabilities exceeding in any year the income and revenue provided for that year.[39] However, a California city would not incur a liability invalid under the constitutional provisions if it entered into a contract to pay for services as and when rendered from time to time in the future. In this case the obligation to pay rental and power charges could not be said to have been incurred until the rental had accrued and the power had been received. Therefore the City of Los Angeles had not incurred a present liability upon the execution of the contracts, and the only effect was to require the appropriation

[36] House Hearings before the Subcommittee of the House Committee on Appropriations, 71st Congress, 2d Session, *op. cit.* Compare p. 1128 with p. 1205, pp. 1132–34 with pp. 1209–12, and p. 1136 with pp. 1213–14.

[37] *Ibid.* Compare pp. 1142–45 with pp. 1220–22.

[38] 36 Opinions of the Attorney General 270, 281.

[39] *Constitution of the State of California,* Art. 11, Sec. 18.

in each annual budget of sufficient funds from the water and power revenues to meet the obligations which would arise in connection with the performance of the contracts. Inasmuch as the Secretary of the Interior was clearly of the opinion that such funds would be available, the Attorney General found no reason to doubt the capacity of the City of Los Angeles to contract. In addition, the lease contract had been formally approved by the board of directors of the Southern California Edison Company and there was no question, therefore, as to the binding effect of the contract upon the corporation. Since the lease contract with the City of Los Angeles and the Southern California Edison Company would provide sufficient funds to comply with the Boulder Canyon Project Act, the Attorney General did not think it necessary to render an opinion concerning the Metropolitan Water District contract. He said that under the circumstances it made little difference whether the aqueduct was built before the power plant or vice versa.[40] Yet he pointed out that, if the aqueduct financing were construed as a prerequisite, the reservation of power for the district was within the authority of the Secretary of the Interior under the provisions of the Boulder Canyon Project Act.

In spite of the opinion of the Attorney General, Arizona did not give up the struggle; and later she filed her objections with the Comptroller General, who rendered his opinion on October 10, 1930.[41] He concurred with the opinion of the Attorney General and decided that the contentions and arguments made on behalf of Arizona were based primarily upon the future possibility that the City of Los Angeles or one of its departments might repudiate the obligations imposed by the contract. He stated that such questions were not within the province of his office and decided, therefore, that the case presented no grounds upon which to justify the withholding of his approval to the appropriation of funds for the Boulder Canyon Project. Subsequently Arizona sought an injunction in the Supreme Court, and brought suit against the other Colorado River Basin states and the Secretary of the Interior. This

[40] 36 Opinions of the Attorney General 270, 282.

[41] Wilbur and Ely, *op. cit.*, pp. 655–62.

move also met with failure (as previously discussed), and the last legal obstacle to the construction of the Boulder Canyon Project was overcome.

Until 1937 no further objections to the provisions of the revenue contracts were voiced; but in that year several revisions were requested, the most important of which was proposed by the Metropolitan Water District. Construction of the dam and power plant had proceeded much more rapidly than had been planned, and the district realized that it would not be ready to take the power for which it had contracted. On May 11, 1937, the Secretary of the Interior, Harold L. Ickes, had notified the City of Los Angeles and the other municipalities that the Bureau of Reclamation would be ready to begin the delivery of Boulder Canyon power to them on June 1.[42] By the terms of the contract, the district would have to begin paying for its power one year later, or on June 1, 1938;[43] but since the aqueduct would not be completed at that time, the district would have no use for the power until a later period. In addition to an extension of time a rate revision was also requested on the ground that southern California was under a competitive disadvantage as compared with other regions served by federal power, particularly in the Northwest where low-cost Bonneville energy would offer a greater inducement to industries to locate in that region than southern California could offer.[44] The Metropolitan Water District based its appeal upon six contentions: (1) It claimed that the power rates of federal projects should be based upon capital and operating costs attributable to power generation only; accordingly, it asked that the $25,000,000 allocated to flood control be removed from the power-rate base of 1.63 mills and that at least $5,000,000 additional be removed and allocated to improvement of navigation. (2) It recommended the operation of the power plant by the federal government alone rather than by the city and the company, on the ground that one agency could operate the power plant with greater efficiency and more economy than two, and estimated that the annual sav-

[42] *The Reclamation Era,* Vol. XXVII, No. 6 (June 1937), p. 122.

[43] Wilbur and Ely, *op. cit.,* p. 164.

[44] *Engineering News-Record,* Vol. CXIX, No. 11 (September 9, 1937), p. 430.

ings gained by this change would be from $220,000 to $400,000. (3) Since power would be ready for delivery to the district ahead of schedule, the district asked that its obligation to take energy be deferred until its aqueduct was completed and the power could be used for pumping. This would mean that the district's obligation would be postponed from June 1, 1938, probably to some time in 1940, with a consequent large loss in revenues to the United States. (4) The district stated that the cost of the power machinery and equipment should be amortized in fifty years rather than in the ten provided in the contract, since the shorter period would make the cost of power unduly high during the initial, critical, load-building period. The extension of the amortization period would mean an annual saving in operating cost of $485,000 during the first ten years of the aqueduct's operation, and the period would then correspond with the length of the contract, thus simplifying the accounting procedure. (5) The four-year load-building period included in the contract was not considered long enough,[45] since the need for power for pumping purposes was not expected to reach its maximum until after a much longer time had elapsed. Hence the district requested that the load-building period be extended to fifteen years, with payments arranged over the entire fifty-year life of the contract so that the ultimate return to the federal government would be the same. (6) Because the 4 per cent interest rate was higher than the actual cost of money to the government, the district asked that the rate be reduced to 3 per cent, thus permitting a further reduction in the power rates.[46]

The City of Los Angeles also expressed an objection to the provision in the lease contract which designated Los Angeles as the generating agency for all state power.[47] The state of Nevada had asked for a comparatively small amount of energy, but the city was not willing to supply the power from its own generating units because of the increased burden and the consequent dan-

[45] Wilbur and Ely, *op. cit.*, pp. 128–29, 166.

[46] *Engineering News-Record*, Vol. CXIX, No. 11 (September 9, 1937), pp. 430–31.

[47] Wilbur and Ely, *op. cit.*, pp. 129–31.

ger to the city's service. It was not practicable to install a large generating unit solely to supply the relatively small needs of the state, and a modification of the arrangement for the generation and delivery of power was requested.[48]

In order to clarify the points involved in the requests for revision the Secretary of the Interior called a meeting to discuss the proposed modification of the Boulder Canyon power contracts. Since any concession granted to one contractor would become effective as to all of the others, the district and the city received support from all of the purchasers of Boulder power in their attempt to lower rates and revise generating arrangements. As a result of this meeting, California and Nevada delegations, on May 21, 1937, requested the House Committee on Rivers and Harbors to add to the pending Bonneville Dam bill a rider permitting the revision of the Boulder Canyon energy contracts.[49] The Boulder Canyon power contracts had nothing to do with the Bonneville Project, but the rider was suggested to bring the proposition before Congress quickly, whereas a separate bill dealing only with the Boulder Canyon situation would have had little chance of enactment during that session. The proposed amendment would reduce the interest rate charged on Boulder Canyon Project indebtedness from 4 per cent to 3 per cent, would relieve the power contractors of interest and amortization charges on the $25,000,000 allocated to flood control until after other features had been paid for, and would permit Arizona and Nevada, at the option of their legislatures, to accept a fixed annual payment of $300,000 each in place of the 18¾ per cent of potential surplus provided for in the act. The charges imposed for the use of falling water would be reduced to compensate for the reduction thus effected. The amendment omitted all mention of a change in generating arrangements, however, probably in order to avoid the unwelcome criticism that the contractors were attempting to escape all of the obligations they had so willingly assumed in order to speed the construction of the project and

[48] *The Reclamation Era,* Vol. XXVII, No. 5 (May 1937), p. 96.

[49] *Engineering News-Record,* Vol. CXVIII, No. 21 (May 27, 1937), p. 793.

to gain a share of the energy produced. The amendment was approved by the Rivers and Harbors Committee, by the representatives of California and Nevada, and by the contractors for 90 per cent of the power; and it was said that the terms were acceptable also to the state of Arizona.[50] Yet when the bill came before the House of Representatives and the Senate, the Boulder Canyon amendment was promptly thrown out in both houses.[51]

No revision was made in the Boulder Canyon power contracts until July 14, 1938. At that time a new contract was signed with the Metropolitan Water District which relieved the district of the obligation to begin taking power on June 1, 1938. The date was extended to June 1, 1940, and the district was not required to take the full 36 per cent of the total firm power available until 1955. The district was not relieved of any payments under the new contract, however. The difference during each month between what the district would have been required to pay under the original contract and what it is now required to pay was to be charged as a debt of the district with interest at 4 per cent. The amount of the debt was to be reduced by any proceeds received from the sale of power which the district had contracted for but had not taken, and the retirement of the debt was to begin in 1955.[52] The original contract had contained a provision giving the Secretary of the Interior the authority to dispose of the district's unused power if the city and other major purchasers failed to exercise their options on it. Under this provision a contract with the Needles Gas and Electric Company was completed on January 28, 1938, for 20,000,000 kilowatt-hours annually; and later another contract for 50,000,000 kilowatt-hours annually was entered into with the Citizens Utilities Company, which operates the power system serving Kingman County, Arizona. These contracts were both to become effective on June 1, 1938, were to expire on December 31, 1954, and would account for revenues of $32,600 and $81,500 a year, respectively. The companies promised to

[50] *Engineering News-Record,* Vol. CXVIII, No. 21 (May 27, 1937), p. 793.

[51] *Congressional Record,* 75th Congress, 1st Session, Vol. LXXXI, Part 7 (1937), pp. 7614–22, 7646–47.

[52] *Engineering News-Record,* Vol. CXXI, No. 3 (July 21, 1938), p. 68.

pay their proportionate share of the cost of the generators installed to serve the Metropolitan Water District and to take their allotment of the energy at the regular price of 1.63 mills per kilowatt-hour. The energy payments were to be made to the federal treasury; but they would be credited against the obligation of the Metropolitan Water District, thereby serving to reduce the charges made against the district.[53]

Contrary to the arrangement thus made with the district, some of the other purchasers of power expressed a desire to begin taking power before their contracts became operative, and at least one interim contract for three years was signed on July 18, 1937, with one of the smaller companies to begin taking power immediately, instead of in 1940, the effective date of the original contract.[54] It was estimated that this new contract would add an unexpected $750,000 to the revenues of the project. Los Angeles and the other municipalities were already taking power and were making advanced payments on the machinery. By July 1, 1937, some $1,200,000 had been paid into the federal treasury, and repayment of the cost of the project was well on its way.[55] In 1939 and 1940 revisions of the plan for the handling of the power and of the factors to be considered in the determination of rates were suggested again, but no actual changes were made.[56] In all cases except that of the Metropolitan Water District, the financial plan of the project seemed to be developing according to the forecasts made when the contracts were signed in 1930; and it appears now that the district will not be forced to bear an unreasonable loss. Although the United States is holding the district to its contract as revised in 1938, it is likely that further supplementary contracts will be made which will relieve the district of a large part of its burden but which will still provide the federal government with the anticipated revenues necessary to meet the requirements of the Boulder Canyon Project Act.

[53] *The Reclamation Era,* Vol. XXVIII, No. 4 (April 1938), pp. 68–69.

[54] *Engineering News-Record,* Vol. CXIX, No. 4 (July 22, 1937), p. 126.

[55] *Ibid.*

[56] *Ibid.,* Vol. CXXII, No. 22 (June 1, 1939), p. 739, and No. 23 (June 8, 1939), p. 749; Vol. CXXIV, No. 2 (January 11, 1940), p. 59.

CONTRACTS FOR WATER

DETERMINATION OF RATES

Although the sale of electric energy would provide by far the largest revenue from the project, the sale of stored water was also expected to develop into an important source of income. The Boulder Canyon Project Act authorized the Secretary of the Interior to make contracts for the delivery of water stored by the dam and provided that no person should acquire a right to the use of such waters except by contract. The act also provided that there was to be no charge made for water delivered to the Imperial or Coachella valleys. The chief revenue to be derived from the storage of water, therefore, would come from the sale of the water to the Metropolitan Water District of Southern California. It was originally estimated that the district would require at least 750,000 acre-feet of water per annum from the Colorado River during the first ten-year period of use in order to replenish underground levels upon which domestic consumption depended; and it was understood that the district might desire to contract for 1,000,000 acre-feet per annum.[57]

The first studies made to determine the proper rate to be charged for Boulder Canyon water proceeded on the assumption that the rate must be sufficient to pay the cost of all features of the project directly involved in providing the storage. Twenty-six million acre-feet was taken as the capacity of the reservoir, and the studies made by the Bureau of Reclamation indicated that with such storage possibilities approximately 10,000,000 acre-feet of additional water per year could be diverted from the river under existing upstream development and 5,000,000 acre-feet under conditions existing in the distant future. The cost of the dam was estimated at $98,000,000 and the annual cost of operation and maintenance, including depreciation, at $291,000. On this basis, the annual charge to provide for repayment, operation and maintenance, and depreciation in fifty years with interest at 4 per cent would amount to about $4,857,000. It was assumed that 5,000,000 acre-feet of

[57] Wilbur and Ely, *op. cit.*, p. 504.

the reservoir would be allocated to silt storage, and 4,400,000 acre-feet to dead storage to serve as a minimum head for power. The active storage would be, therefore, 26,000,000 — (5,000,000 + 4,400,000), or 16,600,000 acre-feet, and the annual amount chargeable to this storage would be (16,600,000 ÷ 26,000,000) × $4,857,000, or about $3,100,000. If 10,000,000 acre-feet were taken as the figure for additional water which could be diverted annually, the cost of the additional water would be 31 cents per acre-foot ($3,100,000 ÷ 10,000,000). If 5,000,000 acre-feet, the future figure, were used, the cost would be 62 cents per acre-foot. It was suggested that a compromise figure of 50 cents per acre-foot would be a fair price to charge. On the basis of an annual consumption of 750,000 acre-feet, the revenue which would be derived from the Metropolitan Water District would be $375,000 per annum (750,000 × $0.50), and this was the water-revenue figure used in determining the rates for power.[58]

During the negotiations with the Metropolitan Water District, however, the construction-cost aspects of the water-rate problem were abandoned and the price was set at 25 cents per acre-foot. This figure was determined on the basis of the cost of delivery and on the district's ability to pay. It was reached by an agreement with the Metropolitan Water District and was not a price fixed by the Department of the Interior and then offered to bidders as in the case of the power rates.[59]

The abandonment of the cost-of-construction basis for the determination of water rates was not so serious as might at first appear. Power and stored water were joint products of a multiple-purpose project, and, as in the case of secondary energy, the project would benefit more by selling stored water at a price which would cover its separable costs and contribute something to the joint costs than by not selling it at all. In addition, the costs of the dam were taken into consideration in the determination of power rates, and each acre-foot of water which passed through the power wheels would already have earned

[58] See Table VII, p. 145, above.

[59] Letter from the Department of the Interior, Bureau of Reclamation, May 9, 1938.

an energy toll of about 75 cents. If the water were to be diverted directly from the reservoir above the dam, therefore, its price would be $1.00 per acre-foot; but diversion below the dam, as in the case of the Metropolitan Water District, would take into account the 75 cents charge already earned and would be burdened with only 25 cents per acre-foot as the fair rate for water which had passed around the dam. It was questionable whether any charge at all for water should be imposed, since the water, if not diverted by the district, would flow unused into Mexico and probably on into the Gulf. Still the district was willing to pay the charge because of the great need of Los Angeles and other cities near by for water; but in view of the immense cost of the aqueduct and the heavy annual pumping charges involved, the 25-cent rate was considered the maximum that the district could pay.

DRAFTING OF THE WATER CONTRACTS

The use of water in California from the Colorado River falls roughly under two headings: (1) water for domestic use by the coastal-plain cities, and (2) water for agricultural use. This difference in interest was recognized by the prospective applicants for the stored water; and on February 21, 1930, an agreement was reached whereby California's share of the water was divided between the domestic and agricultural uses.[60] According to this agreement the agricultural groups, which included the Coachella Valley County Water District, the Imperial Irrigation District, and the Palo Verde Irrigation District, were to receive 3,850,000 acre-feet per annum and the Metropolitan Water District 550,000 acre-feet, a yearly total of 4,400,000 acre-feet which corresponded with the limitation provided in the Boulder Canyon Project Act. If more water were available for California use, the district was to be granted the first 550,000 acre-feet per annum of the excess and all of the remainder was to go to the agricultural groups.[61] The maximum amount of water allocated to the Metropolitan Water District

[60] Wilbur and Ely, *op. cit.*, p. 31.
[61] *Ibid.*, p. 549.

was, therefore, 1,100,000 acre-feet per annum; and it was essential that a contract for the water be executed to enable the district to utilize the electric energy previously allocated to it. Accordingly, a water-delivery contract was completed with the district on April 24, 1930, which preceded by two days the execution of its power contract. Subsequently the district's water contract was amended to concur with a further agreement among the California claimants for Colorado River water.[62] On February 10, 1933, the district signed still another contract with the United States which provided for the construction and operation of Parker Dam to serve as a diversion point for the district's share of the water.[63] This dam was actually a part of the Colorado River Aqueduct project, but it was so closely related to the Boulder Canyon Project that from the district's point of view the third contract was merely another step in the same development.

When the Metropolitan Water District of Southern California was first being discussed, it was suggested that all of the coastal-plain cities from Los Angeles to San Diego should join together in forming one large water district; but further investigations had indicated that a more practical solution to the problem would be to supply the Los Angeles area and the San Diego area through separate systems.[64] When the Metropolitan Water District was formed, therefore, it was decided to build an aqueduct from the river to the coast to supply the Los Angeles vicinity only; and the City of San Diego was left to its own devices to secure a share of the Colorado River water. It was suggested that the City and County of San Diego, acting through the city or a new metropolitan district, should build an aqueduct which would tap Colorado River water from a point near the end of the Imperial section of the All-American Canal. The All-American Canal contract provided that the canal could be used by other agencies if they entered into contracts with the United States to contribute toward the cost of the works;[65] but so far

[62] Wilbur and Ely, *op. cit.*, p. 31. [63] *Ibid.*, p. 311.

[64] California Colorado River Commission, *Colorado River and the Boulder Canyon Project* (1931), pp. 203–4.

[65] Wilbur and Ely, *op. cit.*, pp. 38, 331–32.

no contract of this type has been executed. San Diego was not confined to the use of the All-American Canal as a method to secure water from the Colorado River, since Section Five of the Boulder Canyon Project Act provided that the city could contract with the United States to divert water from the river, and could supply its own means of transporting the water from the diversion point to the coast.[66] Whatever method might be selected, it would probably be some time before San Diego would be in a position to divert Colorado River water; and for this reason the Secretary of the Interior approved a separate water contract with the city on February 7, 1933, which insured San Diego of its water privileges for a definite period. This contract was executed on February 15, 1933;[67] but since no water has been diverted to San Diego, it has never been in effect. On May 16, 1934, a further contract was negotiated and approved as to form by the Department of the Interior which would provide capacity in the All-American Canal for use by San Diego; but this contract also has never been in force.[68]

The agreement of February 21, 1930, had divided California's share of the water between the coastal-plain and the agricultural groups but had made no definite allocation to the members of the agricultural group. During the negotiation of the various contracts, it became apparent that a more specific division among the applicants would be necessary in order to avoid future disputes arising from conflicting interests. Accordingly, on November 5, 1930, the Secretary of the Interior requested the Imperial Irrigation District and all other California agencies which might contract with the United States to co-operate in effecting an allocation which they could recommend to the Secretary of the Interior. A series of conferences was held, and by August 1931 most of the differences among the various agencies had been eliminated. Representatives of the Department of the Interior were then asked to participate,

[66] *United States Statutes at Large,* Vol. XLV, Part 1, p. 1060.

[67] Letter from the Department of the Interior, Bureau of Reclamation, June 30, 1938.

[68] *Annual Report of the Secretary of the Interior for the Fiscal Year Ended June 30, 1934,* p. 31.

and on August 18, 1931, an agreement was reached which was known as the Seven-Party Water Agreement.[69]

The parties to the agreement were the Palo Verde Irrigation District, the Imperial Irrigation District, the Coachella Valley County Water District, the Metropolitan Water District of Southern California, the City of Los Angeles, the City of San Diego, and the County of San Diego. The agreement was ratified by all of its members; but the Palo Verde Irrigation District attached the condition to its ratification that if the agreement were superseded the right should be reserved to the Secretary of the Interior to contract with the Palo Verde District according to the new determination, provided that the Los Angeles and San Diego priorities should not be disturbed. The reason for this condition was that Palo Verde wished to reserve the right to contract in the future according to any judicial determination which might establish in Palo Verde rights other than those provided by the Seven-Party Agreement, especially rights to use the water outside of the district, where many acres of irrigable land might be reclaimed. This condition was subsequently approved by the other parties; and on September 28, 1931, the Secretary of the Interior issued the amended regulations for contracts for the storage of water which incorporated the Seven-Party Water Agreement with Palo Verde's reservation. Thus the agreement was adopted as a uniform water-allocation clause to be utilized in all California water contracts; and the Metropolitan Water District contract of April 24, 1930, was amended to concur with the new regulations.[70] Under these regulations a contract was signed on December 1, 1932, with the Imperial Irrigation District for the diversion of water and for the construction of the All-American Canal development; and the proposed contracts with San Diego and with the Palo Verde Irrigation District were approved by the Secretary of the Interior on April 7, 1933.[71] The Secretary of the Interior also promulgated regulations on February 7, 1933, for the

[69] Wilbur and Ely, *op. cit.*, pp. 32, 553. For a detailed description of the priorities established under the Seven-Party Water Agreement, see *ibid.*, pp. 557–59.

[70] *Ibid.*, pp. 35, 39, 293–95.

[71] *Ibid.*, pp. 321, 347, 359, 371.

delivery of water to Arizona; but the state did not approve the regulations, and no contracts with Arizona have been executed.[72]

LEGAL ASPECTS AND CONTROVERSIAL POINTS

Arizona objected to the water contracts as she had objected to every other phase of the Boulder Canyon Project; but this time she was not alone in her opposition. Actually dissension appeared first among the California contractors for water for agricultural use, although great care had been taken in the drafting of their contracts. Palo Verde's objections to the water allocation had been quickly overcome through the addition of a reservation to the district's ratification of the Seven-Party Water Agreement; but the objections raised against the All-American Canal contract presented a more difficult problem. The drafting of the contract had not been completed until after the conclusion of the Seven-Party Water Agreement; but, in spite of this precaution, controversial points developed. Objections were filed against execution of the contract, and on October 22, 1931, a hearing was held to determine the validity of these objections. The contract, as originally drawn, had provided for a unified canal system to serve both the Imperial and the Coachella valleys. The federal government had undertaken to construct the Imperial Dam at a point about five miles above Laguna Dam, and to build a canal which would connect this dam with the two valleys. The Coachella Valley Land Owners Association objected to the inclusion of their lands in the Imperial Irrigation District and requested a separate contract. On the other hand, the Water Rights Protective Association of Imperial Valley objected to the inclusion of the Coachella lands on the basis that the sufficiency of Imperial's water would be impaired thereby. The Yuma Water Users' Association and the First Yuma Mesa Unit Holders' Association asked that the diversion and carriage works be operated by the United States instead of by the district in order to insure the Yuma vicinity its share of the water and to allow it a greater

[72] Letter from the Department of the Interior, Bureau of Reclamation, June 30, 1938. For copies of the water contracts and an analysis of their contents, see Wilbur and Ely, *op. cit.*, pp. 68–96, 289–378.

share of the power developed along the canal. The Palo Verde Irrigation District objected to the limitation incorporated in the contract which would prevent that district from assigning water to lands outside of its boundaries from which no return flow could be expected for the benefit of the allottees farther down the river. Palo Verde also asked that a limit be placed on the height of the Imperial Dam to prevent backwater damage to the district. Other objections involved the consideration of the delivery of water to Mexican lands, of power contracts to guarantee all costs, and of protection of cultivated lands against seepage damage.[73]

On November 4, 1931, the Secretary of the Interior rendered a decision concerning the objections to the All-American Canal contract. He pointed out that the Coachella Valley was not able financially to furnish security for the construction of a separate canal from the Colorado River, although its need for the water was very great. The only feasible plan was to include the Coachella lands with the Imperial Irrigation District and to construct a joint canal. Secretary Wilbur stated that both districts would benefit economically by unified control. The United States reserved the right to measure water at any point along the system; and, since deliveries would depend largely on releases from the dam, there was no reason to believe that the unified control would impair the water supply of either party. Yuma's objections to the operation of the system by the Imperial Irrigation District were disposed of by noting that the Secretary had the authority to take over the operation of the system if the district failed to comply with the terms of the contract, and that a fair share of the power had already been reserved by the United States for the benefit of the Yuma project. The Secretary stated that the Palo Verde Irrigation District was not unduly limited by the requirement that the district's allotment be used within the district, since the condition attached to the ratification of the Seven-Party Water Agreement provided that the restriction should not apply to water that the district might acquire through a later agreement or judicial determination.

[73] Wilbur and Ely, *op. cit.*, pp. 36, 326–27, 365–66, 556–69.

Palo Verde's objection to the height of Imperial Dam was overruled by the opinion of the engineers of the Bureau of Reclamation to the effect that the district would not be damaged by the backwater. The Secretary pointed out that no plan to deliver water to Mexico could be considered, since no treaty had been reached concerning the division of Colorado River water with that country. Since the plan to secure power contracts to guarantee the cost of the canal and appurtenant structures appeared to be impractical, it was disregarded in order to avoid delay. In the Secretary's opinion, the item of seepage damage was one which could be adjusted internally without reference in the contract, and the discussion of this matter was postponed until the time when the actual damage, if any, had occurred.[74]

The decision of the Secretary of the Interior disposed, therefore, of all of the objections to the All-American Canal contract, and the Imperial Irrigation District endeavored to carry out the conditions precedent to the contract which embraced ratification by the electors of the district, inclusion of the Coachella lands, and confirmation by a court decree. On February 11, 1932, the electors of the district voted by a majority of five to one to ratify the contract, and the contract was executed by the officials of the district on March 1, 1932. However, the Coachella landowners decided not to petition for inclusion, and thus the contract could not become effective.[75] The Imperial Irrigation District then attempted to negotiate a new contract with the United States which would not contain a condition requiring the inclusion of Coachella lands. In spite of the adverse decision of the Secretary of the Interior on November 4, 1931, the Coachella landowners requested a separate contract under which their lands would assume the sole responsibility for the cost of the works necessary to water the valley but which would be physically connected with the Imperial system. These requests were followed by months of negotiations.

By December 1932 the Imperial and Coachella valleys were nearing an adjustment of their differences and the new All-

[74] Wilbur and Ely, *op. cit.*, pp. 564–69.

[75] *Thirty-first Annual Report of the Commissioner of Reclamation for the Fiscal Year Ended June 30, 1932*, p. 32.

American Canal contract was ready for execution. The condition requiring the inclusion of the Coachella lands was eliminated, and in its place was substituted a promise on the part of the Imperial Valley that these lands would be included on petition within thirty days after the contract was confirmed by the court. The Coachella District retained its corporate entity and was assured against assessment for other than expenses required by the contract until water was available for delivery within the Coachella unit. It was given the privilege of collecting and paying obligations of its landowners to the Imperial District and to use its own system of taxation.[76] The contract was approved by the State Engineer of California and by the California District Securities Commission. On January 12, 1933, the contract was ratified by the electors of the Imperial District by a majority of seven to one, and it was validated by the court on July 3, 1933.[77] An appeal which was carried to the Supreme Court by the Coachella Valley County Water District was dismissed in February 1934 after this district had settled its controversy with the Imperial Irrigation District; and on October 15, 1934, a contract was negotiated to provide capacity for the use of the Coachella District in the All-American Canal.[78]

Unfortunately the opposition to the water contracts did not end with the conclusion of the All-American Canal contract controversy. Arizona was still determined to employ every means possible to delay the project; and at the same time that she had attacked the lease and energy contracts with the Metropolitan Water District on the ground that they lacked mutuality of obligation she had declared the water delivery contract of April 24, 1930, with the district to be invalid for the same reason. Arizona claimed that the contract contained no provisions stating when deliveries of water would begin and that therefore the Metropolitan Water District was under no obligation to receive or pay for any quantity of water at all. In

[76] *Engineering News-Record,* Vol. CIX, No. 23 (December 8, 1932), p. 696.

[77] *Ibid.,* Vol. CX, No. 6 (February 9, 1933), p. 131.

[78] Letter from the Department of the Interior, Bureau of Reclamation, June 30, 1938.

addition the United States had reserved the right to discontinue delivery of water and had refused to accept liability for damages due to suspensions or reductions in the delivery of water. Under these circumstances, Arizona stated, the supposed water delivery contract with the Metropolitan Water District was not a contract enforceable at law.[79] It was generally admitted, even by the proponents of the project, that the contract was merely an option for ten years on the part of the Metropolitan Water District to take a certain quantity of water if available and that it was probably not a true contract. The Attorney General of the United States had decided, however, that it was unnecessary to consider the legal status of the water contract, since the revenues from the power contracts would fulfill the requirements of the Boulder Canyon Project Act.[80] The discussion of this phase of the legal problem, therefore, was postponed indefinitely, and has never been solved through authoritative decision.[81]

Arizona threatened to base further objections to the contract on the ground that the use of water in Los Angeles and vicinity would be an illegal use of Colorado River water outside the basin of the river and that in any event the price to be charged for the water was far too low. These objections were never presented seriously, since the proponents of the project very quickly pointed out that the arguments would probably react to Arizona's disadvantage. Commissioner Elwood Mead of the Bureau of Reclamation stated that Arizona had never recognized the riparian doctrine of confining the use of the water to the watershed and could not consistently object to extra-watershed use.[82] The Colorado River Compact had provided for extra-basin use of the waters of the Colorado; and even though Arizona had not signed the compact, the courts would undoubtedly

[79] House Hearing before the Subcommittee of House Committee on Appropriations, 71st Congress, 2d Session, *Second Deficiency Appropriation Bill for 1930*, pp. 1175–76.

[80] 36 Opinions of the Attorney General 270, 282 (1930).

[81] Letter from the Department of the Interior, Bureau of Reclamation, June 30, 1938.

[82] House Hearing before the Subcommittee of the House Committee on Appropriations, 71st Congress, 2d Session, *op. cit.*, p. 1165.

follow the rule laid down in the case of *Wyoming* v. *Colorado* in which the Supreme Court held that an objection to a proposed diversion on the ground that it was to another watershed was untenable.[83] Arizona apparently demanded a higher rate than 25 cents per acre-foot to be charged for water in order to increase her own revenues from the project. The Boulder Canyon Act had not exempted Arizona from a water charge as it had the Imperial and Coachella valleys, and it was pointed out that an increased charge to the Metropolitan Water District might lead to an increased charge for water used on Arizona lands.[84] As a matter of fact, the proposed water contract with Arizona exempted the state from water-storage charges; but if the demand for a higher rate had been granted, it is questionable if Arizona could have secured an exemption for herself.

After the Seven-Party Water Agreement had been signed and further water contracts had been executed Arizona protested once more against the method of procedure that was being followed in the promulgation of the California contracts. Arizona stated that, in spite of the provision in the Boulder Canyon Act which limited California's share of the water to 4,400,000 acre-feet per annum, contracts had been made with California agencies for 5,362,000 acre-feet. Arizona claimed, therefore, that the California contracts violated the provisions of the act by 962,000 acre-feet of water and that, if the contracts were enforced, Arizona would suffer a diminution of her share to that extent.[85] It was true that if the allocations of 1,100,000 acre-feet to the Metropolitan Water District, 3,850,000 acre-feet to the Imperial Valley and others, 112,000 acre-feet to San Diego, and 300,000 acre-feet to Palo Verde were added, the total would reach 5,362,000 acre-feet; but it should have been remembered that these allocations were made by various priorities.[86] The Boulder Canyon Project Act provided that California's share of the 7,500,000 acre-feet per annum allotted to the lower basin should be limited to 4,400,000 acre-

[83] 259 U.S. 419, 466–67 (1922).

[84] Wilbur and Ely, *op. cit.*, pp. 376, 607.

[85] Arizona Colorado River Commission, *Report for Period February 2, 1933, to May 3, 1935*, pp. 15–16.

[86] Wilbur and Ely, *op. cit.*, pp. 557–59.

feet; but California was still entitled to a share of the water in excess of 7,500,000 acre-feet if such a surplus should exist. Undoubtedly the later priorities amounting to 962,000 acre-feet were expected to be fulfilled from this surplus water, and if the surplus did not exist only the first priorities would be effective. The California contracts were not drawn to violate the project act, and Arizona's objection on this point appeared to be based upon an erroneous conclusion.

During the last days of the Hoover administration, Secretary Wilbur attempted to safeguard Arizona's interests, and tendered the state a water contract for the 2,800,000 acre-feet per annum allocated under the project act. Some progress was made in amending the proposed contract to satisfy Arizona's demands; but there was not enough time to complete negotiations before Franklin D. Roosevelt became President of the United States on March 4, 1933. Still Arizona stated that she would oppose the construction of all diversion dams in the lower river until a division of the lower-basin water could be agreed upon. One reason that the negotiations had failed was that Arizona claimed the entire additional 1,000,000 acre-feet which had been allocated to the lower basin by the Colorado River Compact and California refused to sanction this claim. Arizona investigated the conditions surrounding the drafting of the compact and came to the conclusion that at the time the intention had been to allocate to Arizona the full 1,000,000 acre-feet.[87] The state gathered testimony to this effect and then filed an original bill in the United States Supreme Court on February 14, 1934, to request a perpetuation of this testimony.[88] The application for permission to file suit was opposed by California and by Nevada, although Utah, Wyoming, Colorado, and New Mexico approved the application on the condition that they be reserved the right to cross-examine. However, the Supreme Court denied the application on the ground that the evidence would not be admissible in court and that the 1,000,000 acre-feet of water could not be a matter of controversy between Arizona and California

[87] Arizona Colorado River Commission, *op. cit.*, p. 28.

[88] In the Supreme Court of the United States, October Term, 1933, *State of Arizona* v. *State of California, et al.*, p. 3.

in view of California's self-imposed limitation under the provisions of the project act.[89] Arizona then followed a different method of procedure and reopened negotiations for a water contract with Mr. Harold Ickes, the new Secretary of the Interior. A contract was drafted which fulfilled Arizona's demands; but it was opposed by the rest of the Colorado River Basin states. The upper basin argued that the contract did not protect its 7,500,000 acre-feet of water, and California and Nevada were dissatisfied in that the contract did not recognize the California priorities. A hearing was held in Washington on December 17, 1934; but no agreement was reached.[90] Secretary Ickes took no action on the proposed contract and ordered the state delegates to compose their differences and to draft a new contract as a substitute for the Arizona proposal. Two conferences were held, one in Phoenix in January 1935 and one in Salt Lake City in February 1935. Some progress was made, but the questions concerning the inclusion of the waters of the Gila and the recognition of upper-basin and California rights remained as obstacles to prevent agreement. On November 25, 1935, Arizona turned once more to the courts to secure a legal determination of her rights and asked the Supreme Court of the United States for permission to file suit against the other Colorado River Basin states for a fair apportionment of the waters of the river. On May 25, 1936, the Supreme Court denied Arizona's request chiefly on the ground that Arizona had failed to name the United States as co-defendant with the states party to the Colorado River Compact.[91] The court said that the equitable share of Arizona in the unappropriated water impounded in the reservoir could not be determined without ascertaining the rights of the United States to dispose of that water. Since the federal government could not be sued without its consent, Arizona's only recourse after this decision seemed to be to secure the enactment of a bill consenting to the suit. No attempt was made to secure the enactment of the necessary legislation,[92]

[89] 292 U.S. 341, 360 (1934).

[90] Arizona Colorado River Commission, *op. cit.*, pp. 24–25.

[91] 298 U.S. 558, 572.

[92] Letter from Department of the Interior, Bureau of Reclamation, June 30, 1938.

but the state did demand a rehearing of its petition. Arizona claimed that the denial of her petition to sue the other states deprived her of property without due process of law; but on October 12, 1936, the Supreme Court denied the plea for a rehearing on the ground that the whole affair was nothing more than a potential controversy, and thus blocked once more Arizona's attempt to drag this phase of the project through the courts.[93]

In addition to her opposition to the water allocation, Arizona objected to the construction of Parker Dam, located partly on her territory. Under the provisions of the contract with the Metropolitan Water District of February 10, 1933, Parker Dam was to be built by the federal government for the district in connection with the Colorado River Aqueduct project.[94] Arizona was to receive certain rights to power and to the diversion of water, but the state maintained that no structures should be built on the river until her water rights had been definitely established. On July 31, 1933, an opinion was rendered by the Attorney General of Arizona that the Metropolitan Water District had no right to build the dam without Arizona's consent;[95] and on several occasions Arizona protested to the Governor of California, the Metropolitan Water District, and the federal government against the construction of the dam. Yet the preliminary work was started, and in the autumn of 1934 the contractors began to build a temporary structure across the river toward the Arizona side. On November 10, 1934, Governor B. B. Moeur of Arizona set up a military zone on the Arizona side of the river and declared martial law within the zone. Work on the Arizona portion of the dam was thus stopped until the Supreme Court of the United States issued a temporary injunction in February 1935 preventing Arizona from interfering with the building of the dam.[96] On April 29, however,

[93] 299 U.S. 618.

[94] Wilbur and Ely, *op. cit.*, pp. 311–14.

[95] Arthur T. La Prade, Attorney General of Arizona, *Two Opinions of the Attorney General Furnished to the Colorado River Commission of Arizona* (1933), pp. 44–46.

[96] 294 U.S. 695.

the Supreme Court decided that the construction of Parker Dam had not been authorized by Congress and that there was no reason for restraining Arizona in her interference with construction.[97] The court said that no dam could be built on a navigable stream without specific authorization by Congress, and no specific authorization for Parker Dam had been made in the Boulder Canyon Project Act or in any other act. In order to overcome this objection, a bill was introduced into Congress to authorize the completion of Parker Dam and a number of other projects, including Grand Coulee Dam. The bill passed the Senate without difficulty but met opposition in the House—not because of the Parker Dam features of the bill, but because of the Grand Coulee and other features which were accused of being land-reclamation developments contrary to present policy. However, on August 30, 1935, the act was approved,[98] and the work on Parker Dam was resumed.

After the Parker Dam litigation and the controversy in 1936 over the water allocation, Arizona abandoned her attempt to delay the project through the courts; but unless she decides that a better course of action is to ratify the Colorado River Compact and to co-operate with the other Colorado River Basin states, the day will probably come when Arizona will present her case again to the Supreme Court in an attempt to show that she has actually suffered damage and that the argument is no longer merely a potential controversy. When that day arrives, the Supreme Court may definitely determine Arizona's rights to the waters of the Colorado, and thus eventually a legal decision may dispose of the problem that years of negotiations have so far failed to solve.

SUMMARY OF THE PROJECT REVENUES

A survey of the revenue contracts of the Boulder Canyon Project shows that by far the greater part of the project's revenue will be derived from the sale of electric energy. It has already been pointed out that the estimated revenue from the sale of

[97] 295 U.S. 174.

[98] *United States Statutes at Large*, Vol. XLIX, Part 1, p. 1040.

firm energy alone will more than repay the cost of the project[99] and the sale of secondary energy will increase the power revenue measurably. Only one of the water contracts, the contract of April 24, 1930, for the delivery of water to the Metropolitan Water District of Southern California, is likely to provide a revenue from the sale of stored water in the near future. No revenue can possibly arise from the San Diego water contract for a number of years, because the city is not prepared to divert water from the Colorado, and the charges can be imposed only upon actual diversions. Revenues from the Palo Verde water contract are very indefinite, since provision was made for a possible toll on only the extra-basin diversions if such diversions should be sanctioned by the courts in some future decision. The financial plan of the All-American Canal phase of the project was based upon assessments levied under the Reclamation Law against the lands benefited and not upon a charge for the quantity of water diverted. The other water contracts were either for construction or for the diversion of water for which no charge could be made under the provisions of the Boulder Canyon Project Act or the contracts themselves. Therefore the revenue to be derived from the sale of water at 25 cents per acre-foot was confined to the Metropolitan Water District and was estimated at only $12,353,000 over the fifty-year repayment period. Thus the water revenue may be considered as merely an additional factor of safety which will hasten repayments and increase the amounts available for Arizona and Nevada and the surplus fund. Even though the estimates of the power revenue should prove to be optimistic, however, there is little reason to believe that the financial requirements of the Boulder Canyon Project Act will not be fulfilled. Unforeseen events may change the entire aspect of the power industry and of agriculture in the Southwest, but at the present time the financial structure of the Boulder Canyon Project appears to be sufficiently secure to allay any doubt as to its soundness.

[99] Table X, p. 155, above.

TABLE XI.—POWER AND WATER REVENUES OF THE BOULDER CANYON PROJECT*

Assumptions:

Revenue from 100 per cent of firm energy, plus revenue from sale of water, plus revenue from sale of secondary energy

Machinery investment repaid separately by lessees of power plant within ten years

Repayment period, fifty years

Revenue from sale of firm energy at 1.63 mills per kw-hr:		
City of Los Angeles	$121,310,549	
Metropolitan Water District	118,031,886	
Southern California Edison Company	88,523,915	$327,866,350
Revenue from sale of secondary energy at 0.5 mill per kw-hr:		
City of Los Angeles	$ 12,314,526	
Metropolitan Water District	11,981,700	
Southern California Edison Company	8,986,274	33,282,500
Revenue from sale of water at $0.25 per acre-foot:		
Metropolitan Water District		12,353,500
Total gross revenue		$373,502,350
Distribution of revenue for fifty-year period:		
Operation and maintenance	$ 7,262,557	
Depreciation	8,875,553	
Interest charges on all except $25,000,000 allocated to flood control	108,107,007	
Repayment (exclusive of flood control)	82,674,907	206,920,024
		$166,582,326
Surplus:		
18¾ per cent to Arizona	$ 31,234,186	
18¾ per cent to Nevada	31,234,186	
Interest charges on flood control	12,477,929	
Repayment of flood control	25,000,000	99,946,301
Surplus (available for general development on the Colorado River)		$ 66,636,025

* House Hearing before the Subcommittee of the Committee on Appropriations, 71st Congress, 2d Session, *Second Deficiency Appropriation Bill for 1930*, Statement of Secretary Wilbur. pp. 951–52.

Chapter V

CONSTRUCTION PROBLEMS

With the passage of the Second Deficiency Act of 1930, containing an appropriation for $10,660,000 for the Boulder Canyon Project,[1] actual work on the project could begin. The chief political obstacles had been overcome, the financial success of the project appeared to be guaranteed, and the basic legal arguments were soon to be settled. However, the construction problems still remained unsolved, and it was well known that their solution would require careful planning and efficient organization. Both the construction of the dam and power plant at Black Canyon and the construction of the All-American Canal to the Imperial Valley would be engineering feats of unprecedented magnitude. Every phase of the project, including the related Colorado River Aqueduct project, seemed to assume gigantic proportions. It was found necessary to create new materials and to devise new techniques, and the greatest engineering authorities of the country were called upon to contribute their knowledge to the solution of the problem.

Still, the engineering features were not the only construction difficulties to be considered. It was recognized that most of the work would be done in a rough, wild, sparsely inhabited country with adverse climatic conditions and few resources to contribute toward comfortable, healthful living. It would be necessary to move laborers from other parts of the country to the location of the job, and to provide adequate housing facilities for them. These and other construction problems were met and solved with notable speed by the builders of the Boulder Canyon Project. Their solution attracted world-wide attention, and undoubtedly will long remain an outstanding example of the successful development of a great and difficult project.

[1] *United States Statutes at Large,* Vol. XLVI, Part 1, p. 877.

CONSTRUCTION AT BLACK CANYON

HOOVER (BOULDER) DAM

Selection of the site.—The first investigations of dam sites on the Colorado were made in 1901, 1902, and 1903 when a boat reconnaissance was made from Yuma to a point 26 miles above the Virgin River, and when the stream was mapped from the lower end of Black Canyon to the Arizona-Sonora boundary. In the course of these surveys the Boulder Canyon, Black Canyon, Bullshead, Topock, and Parker dam sites were studied, but no construction was attempted.[2] The disastrous flood of 1905–1907 drew attention from the irrigation possibilities of the river to the flood dangers of the stream, and the scene of the surveys was shifted to the upper division of the drainage basin. During these early investigations, a reservoir site above the canyon section was considered desirable, since the silt load was small there and it would not be necessary to provide a large surplus capacity in the reservoir for silt storage.[3] However, none of the reservoirs above Grand Canyon was found to be large enough to serve as a regulator of the river's flow, and a plan to construct a series of reservoirs on the important tributaries of the upper river was soon discarded as an uneconomic method of development. It was discovered that the development of these sites as storage reservoirs would ruin them for power purposes, the use to which they were naturally best adapted. In addition, these reservoirs would not intercept the brief but torrential floods from fully three-quarters of the Colorado's drainage area; and they were located too far away from the flood areas of the lower basin to be used effectively in regulating the river for flood control. All tributaries of the river below the reservoirs would naturally escape regulation; and since it was in this section of the system that the tremendous quantities of silt entered the river, the problem of the flood menace created by silt would not be solved. It was demonstrated, therefore, that no reservoir or series of reservoirs in the upper

[2] Senate Document No. 186, 70th Congress, 2d Session, *Colorado River Development* (1929), p. 40.

[3] G. E. P. Smith, *The Colorado River and Arizona's Interest in Its Development* (1922), p. 539.

basin could furnish as complete control as storage farther down, and thus the future developments in the upper basin will probably be primarily for power generation rather than for regulation of stream flow in the interests of flood control and irrigation.

In 1918 the investigations of dam and reservoir sites on the lower river were resumed.[4] The most suitable sites to be studied were, in downstream order, Glen Canyon, Diamond Creek, Bridge Canyon, Boulder Canyon, Black Canyon, Bullshead, Topock or Mohave Valley, and Parker. One by one the various choices were discarded until the decision rested between Boulder Canyon and Black Canyon as the site for the first great development on the Colorado River. Glen Canyon, located in Arizona a few miles south of the Utah border, is above a number of important tributaries of the stream, and was found to be too distant from the flood areas and power markets for effective flood control and profitable power distribution. It is poorly situated in relation to transportation facilities, and the local geological formation is of sandstone, which is not favorable for the foundations of the high dam necessary for effective regulation and control of the river.[5] The Diamond Creek and Bridge Canyon sites, both located in the lower end of Grand Canyon, have admirable topographic advantages for power development; but the long, narrow canyon upstream from each limits seriously the capacities of their reservoirs for flood and silt control. The reservoir which might be created by the construction of a dam at Bullshead, 66 miles below Black Canyon and 25 miles north of Needles, California, would be too small for a major flood-control and silt-storage structure; but it may be used advantageously some day as an auxiliary regulating reservoir to supplement the Boulder Canyon installation. The Topock site, located a few miles south of Needles, received very serious consideration, since it is well situated in relation to transportation facilities, the flood and farm areas of the lower basin, and the power markets of southern California.

[4] Ray Lyman Wilbur and Elwood Mead, *The Construction of Hoover Dam* (1933), p. 1.

[5] Senate Document No. 186, 70th Congress, 2d Session, *op. cit.*, pp. 150–76.

However, it too was finally eliminated, since the reservoir would be inadequate and, although comparatively small in capacity, would submerge a large area of improved land. The Parker site, located about 155 miles downstream from Black Canyon and five miles northeast of Parker, was also found to be suitable for the construction of only a relatively low dam and a small reservoir.[6] It is well adapted as a diversion point for water for the Parker Valley and for the Colorado River Aqueduct, and is being developed as such at the present time.

The two remaining sites to receive careful study were the Boulder Canyon site and the Black Canyon site, located about 20 miles apart some 30 miles southeast of Las Vegas, Nevada. Black Canyon was the location finally selected; but an earlier tentative decision to make Boulder Canyon the scene of operations gave that name to the project in the legislation that followed. The decision to build in Black Canyon rather than in Boulder Canyon was a difficult one to make. The two sites were occasionally called the upper and lower Boulder Canyon sites; and from the point of view of location there were no important differences between them as far as the purposes of the project were concerned. Both sites were farther from the power markets of southern California than the Topock site; but both were well within the economic transmission range of electricity to those markets.[7] Both canyons were below most of the tributaries of the river, and would therefore provide complete flood control for the lower basin except for the flash floods of the Williams and Gila rivers. The canyons were near enough to the flood and agricultural areas of the lower basin to permit effective regulation of the river for flood control and irrigation purposes, and a dam at either site would inundate practically the same territory. In either case sufficient silt-storage space would be created to solve that problem for many years, especially in view of the fact that upstream developments would probably decrease the amount of silt brought down to the reservoir. If the dam were built in Black Canyon, the lower of the two sites, Boulder Can-

[6] E. C. La Rue, *Water Power and Flood Control of Colorado River below Green River, Utah* (1925), pp. 37–38, 74, 89, 197, 199.

[7] Senate Document No. 186, 70th Congress, 2d Session, *op. cit.*, p. 206.

yon would be flooded, but other important sites for future developments would not be harmed. Geological conditions were such that a high dam could be built in either canyon, and an extremely large reservoir capacity would be secured in either case. In 1921, Mr. F. L. Ransome, geologist for the United States Geological Survey, reported that the rock in Boulder Canyon was a fine-grained granite of excellent quality[8] and that borings indicated that bedrock would be found from 90 to 158 feet below the low-water surface of the river. In Black Canyon the rock was found to be of volcanic origin. It too was of excellent quality and was termed andesite tuff breccia. Bedrock was found at depths of from 110 to 130 feet, and holes drilled to a depth of 557 feet failed to penetrate other formation.[9] There was no doubt in either case that the rock formations would be able to carry safely the heavy load and abutment thrusts contemplated. Earthquake faults were found in both canyons, but none of them showed evidence of recent movement or was so situated as to endanger a high concrete dam.[10]

The early studies indicated that a dam located in Boulder Canyon would be more economical than one in Black Canyon; but further investigations disproved this conclusion. The Black Canyon site was found to be superior to the Boulder Canyon site in several respects. A construction railroad from Las Vegas could be built more easily and more cheaply to Black Canyon than to Boulder Canyon, and it would pass near available gravel deposits and excellent quarry sites. The layout for a construction camp was better there, since despite the ruggedness of the surrounding country and the depth of the gorge the terrain above the 1,500-foot contour, where the quarries, railway yards, shops, and camps would be located, is open and suitable to such use at a reasonable cost.[11] The canyon is narrower than Boulder

[8] *Twenty-first Annual Report of the Reclamation Service* (1921–1922), p. 119.

[9] *Twenty-second Annual Report of the Reclamation Service* (1922–1923), p. 123.

[10] Senate Document No. 186, 70th Congress, 2d Session, *op. cit.*, pp. 94, 97. In 1937 a number of earthquake shocks were recorded near Black Canyon; but these shocks were believed to have been caused by the filling of the reservoir behind the dam and the increased water load on the earth's crust rather than by any activity along the old faults. See *Engineering News-Record*, Vol. CXVIII, No. 19 (May 13, 1937), p. 723.

[11] Senate Document No. 186, 70th Congress, 2d Session, *op. cit.*, p. 148.

Canyon, its walls are steeper, and the concrete volume for a dam of given height there would be less. At the site of the dam Black Canyon is only from 290 to 370 feet in width at low-water level and from 850 to 970 feet at elevation 1,232, the crest of the dam. Mountains of solid rock extend for miles in each direction, thus providing, as a gift from nature, the strongest possible wings for a dam. The maximum depth to bedrock is less than in Boulder Canyon, and the rock formation is more suitable for use in construction. The Black Canyon rock is not as hard to drill as that of Boulder Canyon, and it will stand better in large tunnel excavations. This statement was later confirmed in a very satisfactory manner when it was found unnecessary to use any timbering for roof support in the four tunnels 56 feet in diameter which were drilled through the canyon walls around the dam site.[12] The rock is well-cemented, tough, and durable, and it resists the attack of weather and erosion exceptionally well. For the same height of dam the Black Canyon site gives a larger reservoir capacity than the Boulder Canyon site, and there is a superior location for a power plant below the dam. For these reasons Black Canyon was finally selected as the site of the development, but the name "Boulder Canyon" was retained as the name of the project.

Criticism of the construction plan.—The selection of the Black Canyon site called for the building of a dam far larger than any previously constructed, and as a result the feasibility of the project was opened to question. It was argued that there were so many unknown and indeterminate factors entering into the construction of such a huge dam that the safety of the structure was doubtful and that the project would actually be a menace to the lower basin. It was said that during construction floods would probably wash away completed works and that errors in calculation would cause frequent delays and expensive rebuilding—the dam might possibly be completed but the cost would probably be double or triple the estimates.[13] If after completion the dam should fail, the flood created would probably

[12] Wilbur and Mead, *op. cit.*, p. 2.

[13] House Report 1657, 69th Congress, 2d Session, *Boulder Canyon Reclamation Project* (1926), Part 2, Minority Views, pp. 1–2, 7–8.

destroy Needles, Topock, Parker, Blythe, and Yuma, would wreck the levees of the Imperial Irrigation District, and would create a permanent channel to the Salton Sea. The opponents of the development stated that the danger of such a catastrophe made it unwise to attempt the project and that some other plan should be followed.

In spite of this terrifying forecast, the technical aspects of the problem never awakened serious doubt in the minds of the engineers. Careful investigations had been made before the plans for the great dam at Black Canyon were approved, and months had been spent in making technical studies for the design of the dam (which involved stress problems of unusual magnitude) and in making many comprehensive studies of a research nature.[14] Long before the Boulder Canyon bill received favorable action by Congress it had been decided definitely that the development was possible from an engineering point of view.

It had also been shown that a gigantic structure would be the most economic solution of the various problems of the lower basin. Opposition arguments to the effect that the dam would be too large, or that a series of lower dams would be a better method of controlling the river, ignored the opinions of eminent engineers and the results of numerous studies on the subject. These points were so clear and definite that apparently the criticisms were meant to delay the construction of the project rather than to bring about a drastic revision in the plans. Obviously a dam intended to provide only for flood control might differ in type and location from one intended to store and divert water solely for irrigation. Similarly, development of the river for power alone would call for a disposition of dams different from that required by development for flood control or irrigation. There are theoretical engineering reasons why flood-control and storage works should be erected farther up the river, and also why storage works should be erected farther down the river. Eventually such developments will probably take place. But the immediate problem was to devise a plan to yield the desired

[14] *Twenty-ninth Annual Report of the Commissioner of Reclamation for the Fiscal Year Ended June 30, 1930*, pp. 15–17.

services quickly and in the most economic manner possible under the circumstances. Such a plan would necessarily require a joint solution of all problems and would therefore represent a compromise among the ideal solutions of each.

A series of small dams would not be an economic method of achieving this result. The fixed charges of construction, such as building a camp and installing machinery, are nearly as much for a small dam as for a large one. Within certain limits, the larger the reservoir, the cheaper per acre-foot is the storage and the cheaper per horsepower is the power development. In addition, a large reservoir is more efficient in effecting flood control. With a comparatively small reservoir (of 10,000,000 acre-feet, for instance) the capacity would diminish rapidly with the accumulations of silt, and nearly all of the space would have to be reserved for flood control. With a very large reservoir (say of 30,000,000 acre-feet) space would be available for silt storage for many years without affecting the efficiency of the reservoir, and water storage for agricultural and domestic use could be accomplished as well as control of the largest floods.[15] A high dam at Black Canyon would create the large reservoir required, and since the reservoir would be narrow and deep with a relatively small surface area, there would be little evaporation loss. The water might be carried back as far as Diamond Creek, where, in the future, a power dam may be built; and below Black Canyon another dam may be constructed at Bullshead. With these three dams, the entire head of the river between the west boundary of Grand Canyon National Park and Parker Dam would be utilized, and none of the power resources of the river would be lost.[16] Therefore, from both the engineering and the economic points of view, the construction of a high dam at Black Canyon appeared to present the best solution to the problem and to indicate the future course of development of the river.

Although a very large reservoir was essential to the success-

[15] Senate Hearings before the Committee on Irrigation and Reclamation, 68th Congress, 2d Session, *Colorado River Basin* (1924), Part 1, statement of Arthur P. Davis, Chief Engineer and General Manager, East Bay Municipal Utility District, Oakland, California, p. 74.

[16] *Ibid.*, statement of Mr. Frank E. Weymouth, pp. 101–2.

ful operation of the project in view of the several purposes involved, the objection was made that the great size of the reservoir would be a constant source of difficulty. It was stated that in the wide expanse of territory that would be flooded the water would find faults in the rock or strata of pervious materials through which it could escape to the surrounding desert. A careful survey of the reservoir site, however, disclosed that this conclusion was unwarranted. The water would be confined by mountains of hard rock or by natural dams of softer, but not particularly pervious material, through which it would have to pass for twenty miles or more before finding an outlet. There is no point at which the escape of water from the reservoir through a relatively thin barrier of pervious material need be feared,[17] and leakage around or under the dam would also be improbable. The rock at the dam site has a degree of elasticity, and instead of cracking under the tremendous pressure caused by the filling of the reservoir it would tend to expand laterally. This expansion would be prevented by the surrounding material, and the resulting horizontal compressive force would help to hold the dam in place rather than to weaken its position in relation to the submerged bedrock and the cliffs of the canyon.[18]

During the survey of the reservoir it was discovered that large beds of salt existed in certain sections which would be submerged by the stored waters. Opponents of the project suggested that the solution of this salt might make the water saline and unfit for agricultural or domestic use. However, the salt was found to be coarsely crystalline; and in most cases it was buried under layers of silt. Thus all of the salt would not go into solution at once, and no appreciable increase in the salinity of the water would occur. Even if all the salt should go into solution at once, the quantity of water would be so large that the salt content would not be sufficient to render the water dangerously saline.[19]

Most of the land within the reservoir was public land; but small areas were privately owned. This difficulty was not a seri-

[17] Senate Document No. 186, 70th Congress, 2d Session, *op. cit.*, p. 92.

[18] *The Reclamation Era*, Vol. XXV, No. 4 (April 1935), p. 84, and No. 7 (July 1935), p. 145.

[19] Senate Document No. 186, 70th Congress, 2d Session, *op. cit.*, pp. 144–46.

ous one, since the right of way could be obtained at a comparatively small cost. The villages of Kaolin and St. Thomas, Nevada, and about six and one-half miles of the Los Angeles & Salt Lake Railroad below Overton, would be submerged. A few farms would be destroyed, and several mines would be flooded. But the land in this section was not very productive from either an agricultural or a mining point of view, and the population in the area which would be submerged was less than one hundred. It was estimated that the total value of the property which would be inundated would not reach $500,000. In 1929 the General Land Office posted notices on the public land in the vicinity of the dam and reservoir site announcing that the land had been withdrawn from settlement or any form of appropriation, and proceedings were begun to settle existing private claims and to secure a clear right of way.[20] In 1930, in the first appropriation for construction of the dam, the sum of $500,000 was set aside for the purchase of private properties in the reservoir site; and before January 1, 1933, 82 contracts representing 158 tracts of land in the reservoir site had been executed by the landowners and the government. Payments amounting to $445,502 had been made to 79 owners for 154 tracts, and the government had acquired 6,287.33 acres of land.[21] By the time the reservoir had begun to fill, no private claims remained to delay the project and no objections based upon the destruction of private property could be sustained.

Preliminary construction.—Before actual construction of the dam was begun the Bureau of Reclamation undertook a program of research to establish beyond question the suitability and sufficiency of the methods of design and construction to be adopted. Many preliminary designs were prepared of various types of dams in order to determine the one which would best fulfill the requirements of the project safely and at a reasonable cost. It was decided that owing to the great height of the dam the massive concrete arch-gravity type, built of homogeneous material and depending upon both gravity and arch action to resist the water

[20] *New Reclamation Era*, Vol. XX, No. 8 (August 1929), p. 119.

[21] *Ibid.*, Vol. XXIV, No. 2 (February 1933), p. 23.

pressure, was best suited to fulfill the conditions.[22] The technical design of the dam was largely a matter of mathematical analysis supplemented by model testing; but many problems of a research nature were involved. Studies were made analyzing stresses in massive arch dams, the effects of internal temperature variations, foundation and abutment deformation, earthquake shock, the spreading of canyon walls due to reservoir water pressure, and many other problems created by the unprecedented size of the project. Without doubt the careful studies carried on before and during the construction of the dam are largely responsible for its successful completion and the $385,000 spent in this preliminary work has been more than repaid by the satisfactory results obtained.[23]

As originally drawn, the plans called for a dam which would raise the water level 557 feet and create a reservoir with a capacity of 26,000,000 acre-feet. When the Colorado River Board of Engineers (Sibert Board) met in Denver, on April 12, 1930, however, it was decided to increase the height of the dam to raise the water level an additional 25 feet.[24] This increase in water level to 582 feet would necessitate the construction of a dam 727 feet in height above bedrock, and would create a reservoir with an estimated capacity of 30,500,000 acre-feet. From the economic point of view the higher dam presented certain definite advantages over the lower dam in the development of power and flood control; and the original plans were changed accordingly. Every reservoir has an ideal height of dam which will yield power or water at a minimum unit price. Cost studies of the Boulder Canyon Project showed that per unit power production and water storage costs would decrease up to a reservoir capacity of 34,000,000 acre-feet. Nevertheless the 30,500,000 mark was finally determined as the most economic size for the reservoir because water stored above that level would encroach upon the Bridge Canyon dam site upstream and hinder its possible future development. When

[22] Senate Document No. 186, 70th Congress, 2d Session, *op. cit.*, p. 110.

[23] *New Reclamation Era*, Vol. XXI, No. 6 (June 1930), p. 113.

[24] *Twenty-ninth Annual Report of the Commissioner of Reclamation for the Fiscal Year Ended June 30, 1930*, p. 17.

finally completed, therefore, the plans called for a dam 727 feet in height above bedrock with a bottom thickness of 660 feet, a top thickness of 45 feet, and a crest length of 1,282 feet, built in the form of an arch of 500-foot radius on the axis. The reservoir created by this huge structure would be 115 miles long, with a maximum depth of 590 feet, an area of 146,500 acres (or 227 square miles), and a capacity of 30,500,000 acre-feet.[25]

One of the first important steps to be taken in the construction of the dam was the provision of both railway and highway transportation facilities from the vicinity of Las Vegas, Nevada, to the dam site. The rail route selected started at Bracken, seven miles south of Las Vegas on the Los Angeles & Salt Lake Railroad, and ran 22.7 miles to the summit between the desert plateau and Black Canyon. From the summit to the dam site is 6.7 miles; but because of the rough country it was necessary to locate a route 10.3 miles long in order to secure suitable grades. The first appropriation for the Boulder Canyon Project, approved on July 3, 1930,[26] included the sum of $2,500,000 for railroad construction; and work on the branch line out of Bracken was begun on September 17 of that year. The first 22.7 miles of railroad (plus 5.2 miles of secondary track) were completed on February 5, 1931, and scheduled train service to Boulder City, which had been located near the summit, was commenced on April 17, 1931. The section of the railroad from the summit to the dam site, known as the United States Construction Railroad, was begun on March 1, 1931, and completed early in September. During the next five years this branch line from Bracken to Black Canyon carried by far the greater part of the tremendous quantities of material and equipment that were used in the construction of the dam. After the completion of all construction in Black Canyon, the railroad will remain in use to serve Boulder City and to carry the supplies and machinery that will be required from time to time by the power plant at the base of the dam. Soon after work on the dam site had begun, Six Companies, Inc., the principal contrac-

[25] Department of the Interior, Bureau of Reclamation, *Boulder Dam* (1937).

[26] *United States Statutes at Large*, Vol. XLVI, Part 1, p. 877.

tor, built an additional 20 miles of standard gauge line from the United States railroad to the Arizona gravel deposits, the screening plant, and other salient points.[27] This line was abandoned, however, when its part in the project construction was completed.

In addition to the provision for railway facilities the first appropriation contained an item of $300,000 for the construction of highways, beginning with three miles of road from a point two miles back of the river to the top of the cliffs. Later a contract was made for the construction of a 24-foot oil-macadam highway seven miles long from Boulder City to the dam site.[28] This highway was built for the transportation of workmen and of small equipment and materials to and from the works by trucks during construction and for the future operation and maintenance of the completed dam and power plant. The highway was completed on July 30, 1931. Later, in the vicinity of the dam site, some eight miles of construction roads were built by Six Companies, Inc. In the fall of 1931 the 21-mile highway from Las Vegas to Boulder City was completed by the state of Nevada and connected with the government highway. The contract for the construction of the dam provided for the extension of this highway by 3,500 feet on the Nevada side of the canyon and by about a mile on the Arizona side with a road along the top of the completed dam to connect the highways of the two states. The dam, therefore, is now acting as a connecting link in an interstate highway leading from Las Vegas, Nevada, to Kingman, Arizona.

Another step preliminary to the building of the dam was the provision of construction power, which had to be available at the dam site before any of the heavy work could begin. It was decided to bring in this energy from the outside rather than to build a temporary power plant at the dam site; and $1,750,000 of the first appropriation was allocated to this purpose. The energy was obtained from power plants of the Nevada-California Power Company and the Southern Sierras Power Company

[27] *The Reclamation Era*, Vol. XXV, No. 1 (January 1935), p. 8.

[28] *Thirtieth Annual Report of the Commissioner of Reclamation for the Fiscal Year Ended June 30, 1931*, p. 3.

located more than 200 miles from the dam site. In order to deliver the electric energy it was necessary to construct a transmission line from the San Bernardino substation through Victorville to Black Canyon. Work was started on December 16, 1930, at Victorville, and the line was tied in at the Boulder substation on April 25, 1931. On June 11 the section of the line from Victorville to San Bernardino was completed, and on June 25 electric power for construction purposes was available at the dam site. Between San Bernardino and Victorville the transmission line crosses the San Bernardino range, and at El Cajon Pass it reaches an altitude of 5,500 feet; from Victorville it follows a northeasterly course across 191 miles of desert wastes and rocky hills where winter conditions are severe. Elaborate precautions were taken to insure continuous service, including the provision of another available circuit in case the main line should be out of service between San Bernardino and Victorville. Now that the dam is completed, the line is being used to transmit power from the Boulder power plant to San Bernardino and into the power system of the Nevada-California Power Company to supply a portion of the light and power to be used in southern California.[29]

Specifications.—On July 7, 1930, Secretary Wilbur directed Dr. Elwood Mead, Commissioner of Reclamation, to begin the construction of the dam immediately. Since it was necessary first to complete the preliminary construction of transportation and power facilities, the early work on the project was confined chiefly to surveys. Many new triangulation points were set, permanent bench marks were established, and the town site for the construction camp was laid out. The area of 50 square miles over which construction activities would extend was much too large to be covered adequately by ordinary surveying methods. Therefore contracts were awarded for ground and aerial photo-topographic surveys and for an aerial photographic mosaic map of the entire area in the vicinity of the dam site.[30] With this basic work finished, construction of the dam could begin as soon as

[29] Wilbur and Mead, *op. cit.*, pp. 24–25.

[30] *The Reclamation Era,* Vol. XXIII, No. 10 (October 1932), p. 172.

the specifications had been drawn and the essential facilities had been provided.

During the latter part of 1930 the Bureau of Reclamation speeded up its preparation of the plan of the dam in order to make the specifications available to the prospective bidders at an earlier date than originally scheduled. The effects of the business depression were rapidly becoming more noticeable, and it was hoped that the construction program could be accelerated to provide relief for the unemployed and the ailing "heavy" industries. As a result of this effort the construction schedule was advanced about six months, and copies of the specifications were made available for study in January 1931.[31]

The major items of the specifications called for the construction of the dam (previously described), four diversion tunnels each 50 feet in diameter (two of which would be used later as spillway conduits and two as penstock tunnels), a powerhouse, four intake towers each 30 feet in diameter, two outlet valve houses located on the canyon walls, two overflow spillways each about 650 feet long and connected to the outer diversion tunnels by inclined shafts from 50 to 70 feet in diameter, and a 150-ton capacity permanent cableway for freight purposes. It was estimated that the contractor would have to take out 857,000 cubic yards of common excavation for the foundations of the dam, powerhouse, and cofferdams, and 400,000 cubic yards of rock for the foundation of the dam. Of the 4,400,000 cubic yards of concrete expected to be used for all of the works, 3,400,000 cubic yards would probably go into the dam, making it by far the largest single mass of concrete poured up to that date.[32]

The first step in the construction of the dam would be to build the four circular diversion tunnels (two on each side of the river) through the cliffs around the dam site. These tunnels were to be lined with concrete three feet thick, and were designed to pass a flow of 200,000 second-feet. They were to measure 50 feet in diameter inside of the lining, and would average about 4,000 feet in length. When completed they would be the greatest

[31] *New Reclamation Era*, Vol. XXI, No. 12 (December 1930), p. 245.

[32] *Ibid.*, Vol. XXII, No. 2 (February 1931), p. 32.

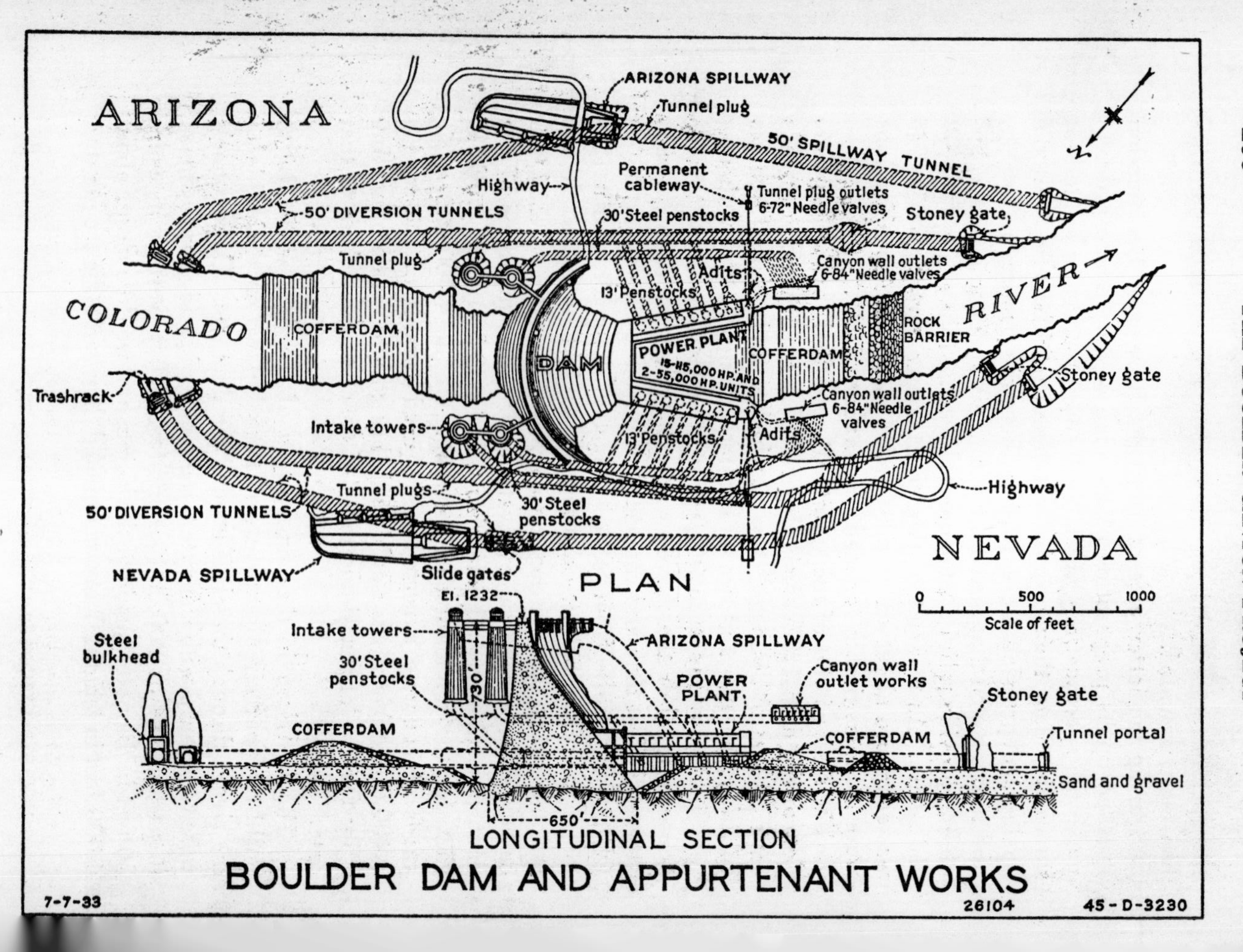

BOULDER DAM AND APPURTENANT WORKS

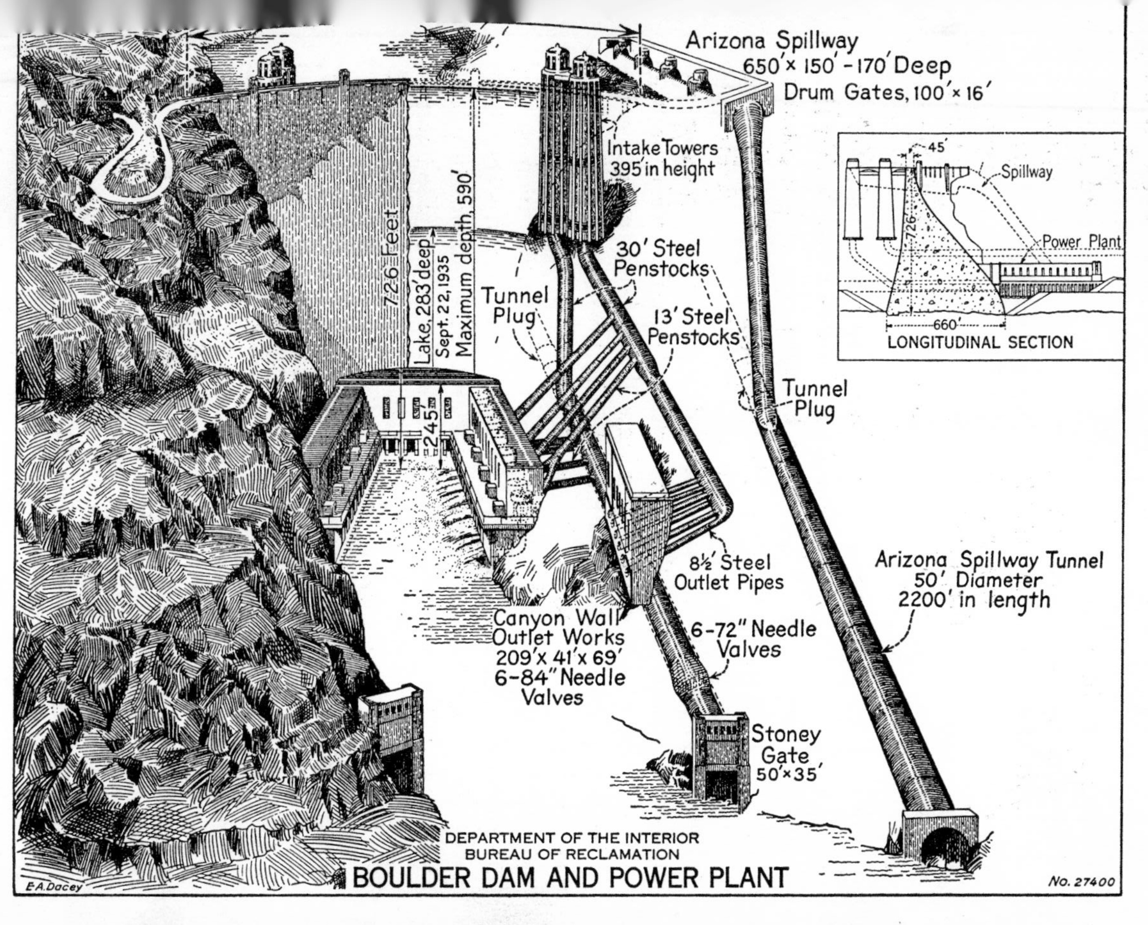

FIGURE 4.—This drawing illustrates the manner in which Hoover (Boulder) Dam works. The Nevada wall of Black Canyon is shown as solid, whereas the Arizona wall is cut away to reveal the intake towers, the spillway, the penstock pipes, and outlet works. Inside the Nevada wall of the canyon a similar set of diversion works has been placed. Principal dimensions are shown.

tunnels in the world with the exception of the Rove Tunnel in France, which has a width of 78 feet 6 inches and a height of 54 feet 4 inches. Both an upstream and a downstream cofferdam were to be built to divert the river from its ancient bed through the tunnels during the construction of the dam, and to turn back the eddy flow from the tunnel outlets. After construction had been completed, the upper cofferdams would be submerged by the rising waters of the reservoir, and the lower cofferdam would be removed to prevent obstruction to the flow from the powerhouse tailrace.

Four intake towers were to be built, two on each side of the canyon, immediately upstream from the dam; and both pairs were to be joined to the dam by a bridge which would connect with the crest of the main structure. Each intake tower was to be the beginning of a separate penstock and outlet system; but, as far as construction was concerned, the four systems were to differ only as between the two systems beginning with the upstream towers and the two beginning with the downstream towers. The water entering the upstream intake towers was to be discharged into inclined shafts each 30 feet in diameter which would connect with the inner diversion tunnels. As soon as the inner tunnels were no longer used for diversion purposes, they were to be completely plugged for 300 feet immediately upstream from the connection with the shaft from the intake towers; and from the connection to the outlet works, they were to contain plate steel pipe 30 feet in diameter. Opposite the powerhouse four tunnels each 18 feet in diameter were to take off from each of the diversion tunnels and extend to the back walls of the power house. A steel penstock 13 feet in diameter was to rest in each of these 18-foot tunnels and connect with one of the turbines in the powerhouse. Beyond the powerhouse the inner diversion tunnels were to contain plug outlet works about 600 feet upstream from the tunnel portals. The water entering the downstream intake towers was to be discharged into tunnels 37 feet in diameter which would extend horizontally downstream through the cliffs. As in the system previously described, these header tunnels were to contain plate steel pipe 30 feet in diameter; and opposite the powerhouse four 18-foot tunnels were to branch off from each side of the river and lead to the powerhouse turbines. On the Arizona side the line at the downstream end of the system was to be split into two pipes 9 feet in diameter to furnish power for two smaller turbines. At the end of each of the header tunnels, six 13-foot bores were to lead outward through the cliffs to the canyon wall outlet works.[33]

Normally the flow of water from the reservoir was to be regulated by the amount required for the turbines of the power plant or by the amount by-passed around the plant through the needle valves in outlet works to satisfy downstream demands or to lower the surface of the reservoir. To provide for exceptional conditions during periods of unusually large floods, however, spillways were to be built on both sides of the river above the dam connecting with the outer diversion tunnels, through which water would reach the river channel again below the dam and the powerhouse. The maximum known flood of the Colorado River, which occurred in 1884, is estimated to have reached a peak discharge of 384,000 second-feet at the dam site.[34] It was considered desirable to limit the reservoir discharge for a flood of this magnitude to 75,000 second-feet; but in view of the comparative shortness of the period of flood records and the consequent uncertainty of the maximum discharge to be expected, the specifications called for a spillway discharge of from 335,000 to 400,000 second-feet. It was considered dangerous ever to permit water to flow over the crest of such a high dam; and, therefore, spillways capable of coping with any possible flood were provided for in the plans. Two spillways, essentially alike in their general features, were to be built, one on each side of the river. They were to be about 650 feet long, 150 feet wide, and 120 feet deep, with the side next to the river formed into an ogee-shaped crest. Each channel would discharge through an inclined shaft 50 feet in diameter and 600 feet long into the outer diversion tunnel on the same side of the river. After diversion activities had ended, each of the tunnels was to be plugged with concrete immediately upstream from its junc-

[33] Department of the Interior, Bureau of Reclamation, *Construction of Boulder Dam* (1936), pp. 28–29, 30–32.

[34] *Engineering News-Record*, Vol. CI, No. 24 (December 13, 1928), p. 888.

tion with the inclined spillway tunnels, and the downstream portion of each outer diversion tunnel would thus become a part of the spillway system.[35]

The project specifications which were sent to the prospective contractors contained several important special conditions including: (1) the assumption by the government of all flood risks after the cofferdams had been accepted, (2) the granting of preference to ex-service men for employment and the prohibition of the use of Mongolian labor, and (3) the reservation of a portion of the Boulder City townsite at a prescribed rental. The time of completion of each of the various units was definitely specified, beginning with the completion of the four diversion tunnels by October 1, 1933, and ending with the completion of the power plant by May 1, 1938. A fixed scale of damages was provided in the event that the contractor should fail to complete any unit by the scheduled date. All materials entering permanently into the construction of the dam were to be furnished by the government with the exception of sand and gravel, and these were to be secured from the government-owned pits. All temporary construction material, such as form lumber, form hardware, temporary timbering in tunnels and shafts, steel liner plates, and all other material not going into permanent construction of the dam were to be furnished by the contractor.[36] The assembling of these materials was in itself a tremendous job, and was one of the first tasks of the successful bidder.

Contracts.—After the specifications of the project were made available, the government advertised for bids on the work for the dam, the powerhouse, and the appurtenant structures, and set March 4, 1931, as the date when the bids would be opened. Six bids were received. The lowest was submitted by Six Companies, Inc., of San Francisco, California, with a bid of $48,890,995, which was about $5,000,000 below its nearest competitor and about $10,000,000 below the third ranking bid.[37] Six Companies, Inc., was composed of six Western con-

[35] Wilbur and Mead, *op. cit.*, pp. 41–42.

[36] *Ibid.*, Appendix, pp. 13–14, 19–22, 23–25, 26*a*, 30–31. See also *Engineering News-Record*, Vol. CV, No. 26 (December 25, 1930), pp. 1014–15.

[37] *New Reclamation Era*, Vol. XXII, No. 4 (April 1931), p. 79.

tracting companies which had joined together to bid for the work. The companies were the W. A. Bechtel and H. J. Kaiser Company of San Francisco, the Utah Construction Company of Ogden, the MacDonald & Kahn Company of San Francisco, the J. F. Shea Company of Portland, the Pacific Bridge Company of Portland, and the Morrison-Knudsen Company of Boise. All of the companies were well and favorably known in general construction work in the Western states. The members of the group had performed all of the various types of construction work required for this undertaking and also had had experience in working together in smaller groups on former contracts. A large and experienced personnel was available which could provide individuals particularly suited to manage and direct various parts of the work; but the organization was so arranged that it would not affect the individual operations of member companies, each of which continued to function independently on other work.

On March 11, 1931, the contract was officially awarded to the Six Companies and the contractors began immediately to move in their plant and equipment and to build their construction camp. The contract was for the construction of the dam, the powerhouse, and the appurtenant structures but did not include the cost of the permanent machinery and materials that would enter into the structures. The government was obligated to provide the cement, reinforcing steel, structural steel, pipe and fittings, conduits, valves, gates and hoists, etc., and from time to time advertised for bids and awarded contracts to supply these materials.[38]

The second largest contract was for furnishing, erecting, and painting plate-steel outlet pipes for the dam, the power plant, and the appurtenant works. The bids were opened on June 15, 1932, and the contract was awarded to the Babcock & Wilcox Company of New York City and Barberton, Ohio, with a low bid of $10,908,000.[39] The company built a fabricating plant at

[38] *Thirtieth Annual Report of the Commissioner of Reclamation for the Fiscal Year Ended June 30, 1931*, p. 2.

[39] Wilbur and Mead, *op. cit.*, p. 67.

Bechtel, one mile from the dam site, where the steel plates, shipped from Barberton, were rolled and the pipe sections fabricated.

Still other contracts were executed for the provision of cement, hydraulic apparatus, etc. Six Companies, therefore, although the principal contractor, was not the only contractor; and a considerable share of the total appropriation for the Black Canyon development was thus paid to other organizations.

Building of the dam.—During 1931 construction of the dam had proceeded rapidly. On May 16 actual construction in Black Canyon was begun when the first blasts were set off at the portals of two adits which had been driven into the abutments of the projected dam, one on each side of the river, to intersect the diversion tunnels. Since March the Six Companies had been busy amassing a vast amount of equipment, building the highways and railroads necessary to its operation beyond those built by the government, constructing the shops, compressor plants, bridges, cableways, concrete-aggregate-screening plant, and concrete-mixing plants, and stripping the canyon walls of loose rock above all tunnel portals. The equipment accumulated included 231 trucks of 30 tons capacity, 2 trucks of 50 tons capacity, 12 electric shovels and draglines, 115 gravel cars of 50 tons capacity, 8 locomotives of 90 tons or larger, a group of 8- and 10-ton cableways, and 5 cableways of 25 tons capacity with self-supporting movable end towers.[40] All electric motors, controls, and other electrical equipment used by the contractor in building the dam were supplied by the Westinghouse Electric and Manufacturing Company under a $2,000,000 contract awarded by Six Companies.[41] A belt conveyor on a suspension bridge was built to carry gravel from the pits eight miles up the river on the Arizona side to the railroad on the Nevada side, and screening and mixing plants capable of manufacturing the huge quantities of concrete required were constructed.[42] The fact that the most efficient equipment and meth-

[40] *The Reclamation Era,* Vol. XXV, No. 1 (January 1935), p. 8.

[41] *New Reclamation Era,* Vol. XXII, No. 6 (June 1931), p. 131.

[42] *The Reclamation Era,* Vol. XXIII, No. 5 (May 1932), p. 94.

ods were used in the work helps to account for the rapid and successful completion of the project.

Of the initial appropriation of $10,660,000, it was estimated that $5,000,000 would be used in the construction of the diversion tunnels, which would require an expenditure of some $18,000,000 before completion.[43] On July 4, 1931, the first direct excavation on the lower portals of the diversion tunnels was begun on the Nevada side, and during 1931 and 1932 the four diversion bores were driven through the canyon walls around the dam site. Lining of these tunnels was started on March 16, 1932, and completed on March 8, 1933.[44] The two diversion tunnels on the Arizona side were lined before those on the Nevada side; and on November 13, 1932, approximately one year ahead of schedule, temporary dams at inlets and outlets of the Arizona tunnels were blasted, allowing the river to flow through the Arizona bores. Within twenty-four hours a dam of muck dumped on both sides of a pile trestle bridge located immediately downstream from the tunnel inlets was built across the river, and the entire flow was forced from its channel. Another temporary dam was completed at a site upstream from the tunnel outlets, and the area between the dams was then pumped dry. Thus the Colorado River, which for centuries had flowed undisturbed through Black Canyon, was turned from its course and diverted around the site of the projected dam.[45]

Excavation to a suitable foundation for the upper cofferdam had been started in September 1932 behind a temporary dike which shut out the river. In October, after the diversion of the river, this work was speeded up, and in March 1933 the upper cofferdam was completed. Work on the lower cofferdam and rock barrier was retarded by the stripping of the canyon walls above the outlet works; but these downstream structures were practically completed by April 15, 1933, although the stripping work continued until February 1934. The cofferdams withstood the spring floods of 1933, and excavation of the foundations of

[43] *New Reclamation Era*, Vol. XXI, No. 8 (August 1930), p. 146.

[44] *The Reclamation Era*, Vol. XXV, No. 1 (January 1935), pp. 8–9.

[45] *Ibid.*, Vol. XXIII, No. 12 (December 1932), p. 198.

the great dam was carried on rapidly. By July 1933 the work was approximately eighteen months ahead of schedule.[46]

Drilling and blasting for the spillway channels was started in February 1932, and excavation was practically completed by April 1, 1933. The first concrete was placed in the weir crest of the Nevada spillway in March 1933, and soon afterward pouring was begun at the Arizona spillway. The placing of the concrete lining in the inclined tunnels was begun in August 1933, on the Nevada side, and was finally completed in October 1934. In March 1934 the work of erecting four 500,000-pound steel drum gates on the crest of each spillway to give automatic control of reservoir overflow during flood periods was inaugurated, and by January 1935 all eight had been assembled. Early in 1935, soon after the erection of the gates, both spillway structures were completed.[47]

Work on the intake tower sites commenced on March 10, 1932, and was practically finished on April 1, 1933. The four huge grooves for the towers cut from the canyon walls required the blasting of some 333,000 cubic yards of rock. Concrete was first poured in the Nevada downstream tower in November 1933, and all four towers were completed in the latter part of 1935. The two cylindrical gates in each tower, which control the flow of water from the reservoir to the power plant or outlet works, were put in place during 1934, and in 1935 the 6,000,000 pounds of trash racks, required to free the water of debris before entering the power system, were installed.[48]

Final approval of the design of the dam had been given on November 19, 1932, by the Colorado River Board, and excavations for the abutments of the main dam structure were commenced at the crest elevation during December. By April 1933 the excavations had been completed down to the bench level, some 600 feet below, and removal of the channel fill to the low point in the bedrock was finished in June. On June 6, 1933, the first concrete was poured in the dam, after twenty-seven months

[46] *Annual Report of the Secretary of the Interior for the Fiscal Year Ended June 30, 1933*, p. 7.

[47] *The Reclamation Era*, Vol. XXV, No. 1 (January 1935), pp. 9–10.

[48] *Ibid.*, pp. 10–11.

of intensive preparatory work,[49] and Six Companies planned to place the entire estimated 3,400,000 cubic yards of mass concrete[50] within the following twenty-three months — nine months less than the time allowed in the first estimate. One million yards of concrete were poured by December 1933, and 2,000,000 yards were in place by June 1934. The 3,000,000 mark was reached in December, and on February 22, 1935, twenty-one months after concreting operations had started, the last bucketful of concrete was placed in the main body of the structure.[51] On March 23 all blocks were brought up to crest elevation, and parapets, utility towers, and elevator towers were erected during the following three months.[52]

The placing of such a tremendous volume of concrete in a single mass involved construction difficulties never before encountered. When concrete sets, a large amount of chemical heat is generated, and as it is dissipated a shrinkage occurs which causes the concrete to crack. In a dam of ordinary size the heat disappears about as fast as the dam is built; but in the case of the Black Canyon structure under ordinary procedures the distance of heat conduction would have been so great that more than one hundred and fifty years would have elapsed before the concrete would have assumed the temperature of the surrounding medium.[53] If the dam had been built without control of the rate of placing concrete and special provisions for its cooling, destructive volume changes would have taken place, resulting in undesirable and possibly dangerous open joints in the mass. To cope with this situation, the dam was designed to have circumferential contraction joints as well as the usual radial contraction joints; and the dam was thus built in a series of 230 vertical columns 25 feet by 60 feet in cross section. None of the columns was carried up more than 30 feet at one time. Moreover, one-inch tubing was placed at about five-foot intervals both

[49] *Engineering News-Record,* Vol. CX, No. 24 (June 15, 1933), p. 789.

[50] The amount of concrete actually poured in the dam itself was 3,240,871 cubic yards. See *Engineering News-Record,* Vol. CXVI, No. 8 (February 20, 1936), p. 299.

[51] *Engineering News-Record,* Vol. CXIV, No. 9 (February 28, 1935), p. 330.

[52] *Ibid.,* Vol. CXV, No. 26 (December 26, 1935), p. 880.

[53] *New Reclamation Era,* Vol. XXII, No. 8 (August 1931), p. 164.

vertically and horizontally in the concrete while it was being poured, and water at a temperature as low as 38° F. was passed through this tubing until the temperature of the surrounding concrete had been lowered to the desired degree. The resulting contraction of the concrete opened spaces between the columns, and these were filled with a water-cement mixture (grout) through half-inch diameter pipe that had been placed as concreting progressed.[54] Cooling was started on August 8, 1933,[55] and was completed in May 1935, about twenty-one months later. Grouting of contraction joints was begun in May 1934 and was in progress for thirteen months until its completion in June 1935.[56] Thus the cooling and contraction which might have taken nearly two hundred years under ordinary circumstances was completed in less than twenty-two months, and the danger of cracking due to heat radiation was avoided.

The Babcock and Wilcox steel plant at Bechtel was completed in the spring of 1933, and the first pipe was finished in April of that year. Since many of the pipes fabricated were 30 feet in diameter with a thickness of 2¾ inches, the plant itself was necessarily a very large affair, and was equipped with unusually powerful machinery. The smaller pipes manufactured were transported by the government from the plant to the Nevada rim of the canyon on the United States Construction Railroad. The larger sections were too heavy for the railroad, and each had to be conveyed on a 200-ton trailer pulled by two tractors. At the canyon rim the pipes were picked up by a giant cableway built under contract by the government and spanning the canyon above the power plant and the four construction adits.[57] This cableway is a permanent part of Black Canyon construction. It is about 1,200 feet in length, has a maximum lift of over 600 feet, operates at a conveying speed of approximately 240 feet per minute, and has a capacity of 150 tons. The first service of the cableway was to lower the fabricated sections

[54] Department of the Interior, Bureau of Reclamation, *Construction of Boulder Dam* (1936), p. 22.

[55] *Annual Report of the Secretary of the Interior for the Fiscal Year Ended June 30, 1934,* p. 29.

[56] *Engineering News-Record,* Vol. CXV, No. 26 (December 26, 1935), p. 880.

[57] *The Reclamation Era,* Vol. XXV, No. 1 (January 1935), p. 11.

of welded steel pipes to the landing platforms on each side of the river; later it was used to transport generating machinery for the power plant, and gates and valves and other miscellaneous machinery for the outlet works. Now it is being run for general maintenance purposes, and handles materials in loads of from one ton to the maximum.[58]

By the end of 1934 all of the smaller pipe sections, such as those leading from the canyon-wall valve houses, had been installed, and the work of placing the large sections in the tunnels leading from the downstream intake towers had begun. Erection of 30-foot penstock header sections was started in the upper Nevada tunnel on July 10 and that of 13-foot diameter penstocks from the same tunnel on September 5. On December 1, 1934, half of the 44,000 tons of pipe required had been fabricated and some 2,500 tons had been installed. By June 1935, 31,000 tons of pipe sections had been produced and over 35 per cent of the work of installation had been finished. In December the 41,000-ton mark was reached in production, of which 33,000 tons had been installed. Seven months later both the fabrication and installation of the pipe sections had been completed.[59] The first penstock was tested successfully on December 26, 1935, and the tailrace of the dam was filled with water for the first time on February 4, 1936. The Babcock and Wilcox Company contract was virtually completed by September of that year, and at that time the 30-foot diameter penstocks, the largest ever built, were tested under an 800-foot head, one and a half times the maximum operating head, and proved their practicability.[60]

The entire construction program of the development at Black Canyon was carried out with unusual smoothness and with very few avoidable delays. There were occasional disputes between the contractors and the workmen; but in every case these disputes were settled quickly and to the satisfaction of both parties to the controversy. On August 7, 1931, a strike was called; but the men

[58] Wilbur and Mead, *op. cit.*, pp. 43–44.

[59] *Annual Report of the Secretary of the Interior for the Fiscal Year Ended June 30, 1936*, p. 64.

[60] *Engineering News-Record*, Vol. CXVII, No. 12 (September 17, 1936), p. 423.

returned to work on August 16. During this early stage of the construction program adequate living facilities for the workmen had not been completed, and the strike may be attributed to the lack of ordinary comforts in living and working conditions. With the assurance of better facilities in the near future and with the coming of cooler weather, the rate of labor turnover dropped considerably.[61]

Early in 1932 the project faced the possibility of serious delay and of large losses to the government due to a proposed cut in the appropriation necessary to carry the project through the year. In an effort to curtail governmental expenditures, the committee on appropriations cut the $10,000,000 figure approved in the budget to $6,000,000. This would have meant the employment of about nine hundred fewer men and might have jeopardized the diversion of the river during the following winter. If diversion had not been effected, it would have been necessary to close down the work for six months, thus precipitating a real calamity for Boulder City and for the construction army encamped there and preventing the orderly delivery of materials to the project. The government, too, would have suffered a distinct loss, since the contract period could have been extended only at an additional cost and with the consequent delay in the date when revenues from the project would begin. The seriousness of the situation was brought to the attention of Congress, and during the summer a relief bill was passed containing an appropriation for additional funds to carry on the project work according to schedule.[62] From that time on no further difficulties were experienced in securing adequate appropriations from Congress to carry on the construction of the project.

During the winter of 1934–35 the dam began to do its work. The storage of water in the reservoir was begun on February 1, 1935, about eighteen months ahead of schedule, when the bulkhead gate of the upstream portal of the Arizona outer diversion tunnel was closed. This event marked the beginning of river

[61] *Engineering News-Record*, Vol. CVII, No. 10 (September 3, 1931), p. 385; also No. 18 (October 29, 1931), p. 707; and No. 21 (November 19, 1931), p. 795.

[62] *Ibid.*, Vol. CIX, No. 3 (July 21, 1932), p. 86.

regulation and flood control with the resulting protection to the flood areas of the lower river and the consequent insurance of an all-year supply of water for irrigation and domestic purposes.[63] By June 30, 1935, the reservoir contained about 3,875,000 acre-feet of water. On that day a flow of 50,300 second-feet recorded at the Bright Angel gauging station was reduced to 14,900 second-feet below the dam.[64] On September 30 the dam was dedicated by President Roosevelt at ceremonies attended by the governors of six states, cabinet members, high officials of the Bureau of Reclamation, and some 12,000 visitors.[65] On February 6, 1936, the lake which was rapidly being formed in the reservoir was officially named Lake Mead, in honor of the late Dr. Elwood Mead who had given so many years of service to the Reclamation Bureau.[66] The dam and the powerhouse were officially accepted by Secretary of the Interior, Harold L. Ickes, on behalf of the United States, on March 1, 1936, about two years and two months ahead of the original schedule. This act terminated the Six Companies' contract, and marked the end of the contract life in eleven days less than five years. After the contract had been awarded on March 11, 1931, several changes were made in the design of the dam and other structures, so that the contractors' gross earnings amounted to about $54,700,000 instead of to the $48,890,995 in their bid. With deductions made by the Bureau of Reclamation for various services, such as the provision of electric power and building space in Boulder City, the actual cash payments to the contractors amounted to approximately $51,950,000.[67] Only one important item in the Six Companies' contract was not completed when the contract was terminated. This uncompleted work was the plugging of one of the diversion tunnels through which it was necessary to by-pass water for irrigation purposes until the powerhouse was placed in operation. This plug was later installed by the Bureau of Reclamation on force account, and thus the chief structure

[63] *Engineering News-Record,* Vol. CXIV, No. 6 (February 7, 1935), p. 232.

[64] *Annual Report of the Secretary of the Interior for the Fiscal Year Ended June 30, 1935,* p. 53.

[65] *Engineering News-Record,* Vol. CXV, No. 14 (October 3, 1935), p. 479.

[66] *The Reclamation Era,* Vol. XXVI, No. 2 (February 1936), p. 35.

[67] *Engineering News-Record,* Vol. CXVI, No. 8 (February 20, 1936), p. 299.

of the Boulder Canyon Project was completed and its routine work begun as contemplated in the construction design.

THE POWER PLANT

During the construction of the dam and its appurtenant structures, the building of the power plant, the second important development in Black Canyon, was begun. The power plant is, of course, an integral part of the water system of the dam, and the construction of the building to house the power machinery was included in the Six Companies contract let on March 11, 1931. The intake towers control the flow of water to the powerhouse and the various outlets, and the inner diversion tunnels are used to bring a part of the water supply to the turbines now that the use of the tunnels for diversion purposes is ended. The specifications called for the location of the powerhouse immediately downstream from the dam and its construction as a U-shaped structure of concrete and structural steel with one wing of the U on each side of the river and with the connecting portion constructed across the downstream toe of the dam.[68] The physical characteristics of the site were excellent for this arrangement, since a rock shelf of sufficient width to accommodate practically all of the building lay at about the elevation of the draft-tube floor on each side of the river. The building, as planned, was to be about 1,650 feet long with 650 feet as the length of each wing and 350 feet as the length of the dam section. It was to be 150 feet above the normal tailrace water surface and 229 feet above the lowest foundation elevation.[69] It was to have a width of 150 feet at the generator floor in each wing, and a maximum width of 157 feet across the central portion. The roof was to cover an area of about four acres and was to be composed of seven laminations 4½ feet thick to resist rocks which might fall from the cliffs above. Beneath the roof were to be about ten acres of floors. Access to the building from the upstream side was to be by means of two elevators in the dam,

[68] Wilbur and Mead, *op. cit.*, Appendix, p. 16.

[69] Department of the Interior, Bureau of Reclamation, *Dams and Control Works* (1938), p. 20.

and a railroad track from a cableway landing below the plant was to be provided for handling heavy equipment.

When finally completed the powerhouse was to contain fifteen 115,000 horsepower units and two of 55,000 horsepower, making a total installed capacity of 1,835,000 horsepower. Eight large units were to be located on the Nevada side of the river and seven large units with the two small units on the Arizona side. It was estimated that when construction was completed in Black Canyon a continuous firm power output of about 663,000 horsepower, or about 4,330,000,000 kilowatt-hours, per year, would be available;[70] but this amount would decrease by about 8,760,000 kilowatt-hours per year each year thereafter owing to the silting of the reservoir and to upstream consumptive uses. It was also estimated that 1,550,000,000 kilowatt-hours per year of secondary power would be generated, but this amount would decrease by 8,600,000 kilowatt-hours each year thereafter.[71] These figures and the figures for the installed capacity are much larger than originally planned, and as a result more power than anticipated was made available for sale on the southern California market. There seems to have been no doubt of the capacity of the market to absorb the greater amount, however, and the original rates were announced for both firm and secondary power even though these rates had been computed on the basis of a smaller installed capacity and a smaller power output.

Construction could not begin on the powerhouse until considerable work had been done on the dam and other structures; but as early as June 1932 plans were being prepared for hydraulic and electrical machinery for the power plant. By January 1934 the foundations for the powerhouse had been excavated and the first concrete was poured for the powerhouse wings. By the end of June, concreting was in progress for the entire lengths of both wings.[72] A year later the walls had been raised to nearly the full height and most of the structural steel

[70] One horsepower = 0.746 kilowatt. Therefore, 663,000 (horsepower) × .764 × 24 (hours) × 365 (days) = 4,332,678,480 kilowatt-hours per year.

[71] *New Reclamation Era,* Vol. XXI, No. 6 (June 1930), p. 114.

[72] *Annual Report of the Secretary of the Interior for the Fiscal Year Ended June 30, 1934,* p. 30.

for the roof had been placed. Installation of power-plant machinery by the government was begun in February of that year, and the draft-tube liners for the first four of the 115,000 horsepower units were installed for the turbines.[73] At that time it was estimated that power would be ready for delivery to the City of Los Angeles by June 1936; but because of delays in the installation of generating equipment, power was not available until the autumn of the year.[74]

Before the powerhouse was completed work had been started on the transmission lines which were to carry Boulder Canyon power to its various points of use. Of the three large contractors for electric energy, the City of Los Angeles was to construct the lines for itself and the smaller communities, the Southern California Edison Company was to be the transmitting agency for the other private companies, and the Metropolitan Water District was to build the lines over which energy would be delivered to the aqueduct for pumping purposes.[75] Actually the first transmission line to be constructed to the dam was the Southern Sierras Company line, which was originally used to transmit power to the project for construction purposes and which is now used to carry energy from the dam to the company's substation at Riverside, California.[76] The second two were the gigantic 287,500–275,000-volt transmission lines, the largest in the world, which were built by the City of Los Angeles at a cost of $22,800,000.[77] Construction of these was started in June 1933 and was completed three years later. Two rows of towers 109 feet high and spaced 800 to 1,000 feet apart were built to carry the conductors a distance of 230 miles from the power plant to El Cajon Pass, and single towers 144 feet in height were erected to carry the two circuits the remaining 40 miles to Los Angeles.[78]

[73] *Annual Report of the Secretary of the Interior for the Fiscal Year Ended June 30, 1935*, pp. 53–54.

[74] *Engineering News-Record*, Vol. CXVI, No. 21 (May 21, 1936), p. 750.

[75] *Ibid.*, Vol. CXIII, No. 22 (November 29, 1934), p. 690.

[76] *The Reclamation Era*, Vol. XXVIII, No. 10 (October 1938), p. 213.

[77] *Ibid.*, Vol. XXIV, No. 4 (April 1933), p. 47.

[78] Department of the Interior, Bureau of Reclamation, *Construction of Boulder Dam* (1936), p. 44.

In addition to these first three, seven other transmission lines were built; but none of them has nearly as high a voltage capacity as the two built by the city. The Metropolitan Water District line, 237 miles in length, was begun in December 1935 and completed on July 26, 1937.[79] The Southern California Edison line, which extends a distance of 238 miles from the dam to the company's receiving station at Chino, California, was started early in 1936, and was completed about July 1, 1938.[80] The Needles Gas and Electric Corporation built a line to Needles, California, which would also supply Searchlight, Nelson, and Eldorado Canyon, Nevada, for mining developments. A line was constructed by the Lincoln County Power District to Pioche, Nevada, and another one was built by the Citizens Utilities Company to Kingman, Chloride, Oatman, Gold Road, and Catherine, Arizona. A relatively small line was built to Boulder City by the Bureau of Reclamation, and a similar line was constructed to Las Vegas by the Southern Nevada Power Company.[81] At the present time an eleventh line is being built by the City of Los Angeles; it is similar to the two already constructed by the city and is expected to be completed in 1940.[82]

After the dam and powerhouse had been officially accepted by the Department of the Interior on March 1, 1936, the Bureau of Reclamation took over the Six Companies' plant and equipment for use without charge until October 1, for the installation of powerhouse machinery. Power production could begin as soon as the installation of one of the several units had been completed; and on September 11 a small 3,500-horsepower generator was put into service to supply current for use at the dam and in Boulder City. This first generator began its work when President Roosevelt pressed a golden key in Washington, D.C., as a part of the ceremonies of the World Power Conference; but large-scale production was delayed until the following month.[83] On October 7, 1936, the first of the 115,000

[79] *Engineering News-Record,* Vol. CXIX, No. 7 (August 12, 1937), p. 282.
[80] *Ibid.,* Vol. CXVII, No. 24 (December 10, 1936), p. 829.
[81] *The Reclamation Era,* Vol. XXVIII, No. 10 (October 1938), p. 213.
[82] *Ibid.,* Vol. XXX, No. 4 (April 1940), p. 108.
[83] *Engineering News-Record,* Vol. CXVII, No. 12 (September 17, 1936), p. 423.

horsepower generators went into service,[84] and on October 9 the first power was transmitted to Los Angeles, although it was not until October 22 that regular service was started.[85] By June 1, 1937, the plant was ready to settle into its routine of power production. Notices were sent to the cities of Los Angeles, Burbank, Glendale, and Pasadena on May 11 that in accordance with the contracts with those cities, 1,250,000,000 kilowatt-hours of energy per year would be available on June 1 at the rate of payment of 1.63 mills per kilowatt-hour[86] and that from that date forward all Boulder Canyon power purchase contracts would be in full force and effect. The contracts for the purchase of Boulder Canyon power negotiated with the utility companies and other agencies, in addition to the four municipalities, contain provisions which make them automatically operative at definite intervals after the date when power is delivered on the Los Angeles contract.[87] The delivery of energy to the Nevada-California Electric Corporation was started on August 16, 1937, and the Lincoln County Power District of Nevada received its first portion of Nevada's power on May 3, 1938. Delivery to the Needles Gas & Electric Company began on September 29 and to the Citizens Utilities Company of Kingman, Arizona, late in October. During 1938 over 1,500,000,000 kilowatt-hours were generated.[88] The Metropolitan Water District took some energy in December 1938 and January 1939 to pump water into the various reservoirs along the Colorado River Aqueduct, and it was expected that the Southern California Edison Company would ask for a share of the power a number of months before June 1, 1940, the date of the normal operation of its contract. It was estimated that the revenues to the government from the sale of the power in 1937 amounted to $1,500,000 and that during the following few years this amount would increase rapidly under the existing contracts.[89] With the fulfill-

[84] *Engineering News-Record,* Vol. CXVII, No. 15 (October 8, 1936), p. 526.

[85] Letter from the Bureau of Reclamation, Washington, D.C., December 1, 1938.

[86] *Engineering News-Record,* Vol. CXVIII, No. 19 (May 20, 1937), p. 756.

[87] Wilbur and Ely, *op. cit.,* pp. 131–32, 166.

[88] *The Reclamation Era,* Vol. XXIX, No. 1 (January 1939), p. 14.

[89] *Ibid.,* Vol. XXVII, No. 7 (July 1937), p. 147.

ment of the conditions necessary to the operation of the contracts the revenues from the development surpassed the expenditures, and the repayment of the cost of the Boulder Canyon Project was begun.

BOULDER CITY

The construction of the great dam, the power plant, and the appurtenant structures required the employment of several thousand men in a rough, desert country some thirty miles from Las Vegas, Nevada, the nearest sizable community. From the beginning it was obvious that it would be necessary to build a large construction camp near the dam site and that unusual precautions would have to be taken to shelter the workmen from the extreme heat of the summer and the cold of the winter, to provide them with a good water supply and adequate sanitary facilities, and to give them recreational opportunities and other conveniences which would provide comfortable living on a heretofore uninhabited section of the desert. To solve these problems a city was built near the dam site which came to be known as Boulder City, and which not only served as a construction camp for the development in Black Canyon but became the permanent home for those employed in the operation of the dam and the power plant. The city was not allowed to grow haphazardly or to develop according to the wishes of those who might want to enter various business enterprises in the new community. Since it was built on public land, it was subject to federal control; and it became one of the most carefully planned cities in the world.

The city was located on the Nevada side of the river about seven miles by highway southwest of the dam site and twenty-five miles southeast of Las Vegas. It is situated on the divide between the river and the plain sloping toward Las Vegas, and has a view of the newly formed reservoir toward the north.[90] It has an elevation of about 2,500 feet above sea level, 1,850 feet above the river, 1,000 feet above the top of the canyon at the dam site, and 470 feet above Las Vegas.[91] This site was selected

[90] *The Reclamation Era,* Vol. XXIII, No. 8 (August 1932), p. 137.

[91] *New Reclamation Era,* Vol. XXII, No. 6 (June 1931), p. 118.

because the 3 per cent slope on the Las Vegas side of the summit rendered construction less difficult than on the steeper and more rugged slope on the river side and because the average summer temperature was found to be lower there than that of any other site considered. The temperatures at Boulder City vary during the year from 20° to 120° above zero, with mean temperatures of 52° in December and 94° in July. The city was located where it has the advantage of the prevailing winds, and the average temperature is seven degrees cooler than at the dam site and two degrees cooler than at Las Vegas.[92] The climate is exceedingly dry, and only occasional desert shrubs were found growing in the sandy, rocky soil before the city was built. With the development of an adequate water supply, however, trees, grass, and flowers were planted and have been found to flourish as long as enough water for their needs reaches them.

It was decided that the federal government should retain ownership of the land on which the city was built and should lease space to those who were to live on it or use it for commercial purposes. The government was never to give up its control over the land, and one of the features of the leases was that they were to continue no longer than the good behavior of the tenants. Bootleggers, gamblers, and other undesirables were not allowed to interfere with the well-being of the workmen; and Boulder City, in spite of its sudden birth and rapid development, did not take on the aspects of a boisterous frontier town. The city was planned by Mr. S. R. De Boer, city planner and landscape architect of Denver, Colorado, and was designed to include the contractor's camp, the government camp, and various business establishments. The city was built in the shape of a triangle, with the apex at the government administration building on the summit of the divide, and the streets, business section, residence section, and parks were carefully laid out. The streets were graded and oil-surfaced or paved with concrete, concrete curbings and sidewalks were constructed, and a street-lighting system was installed. The government constructed a town hall, a school, a garage, a dormitory and guest house, an auditorium,

[92] Wilbur and Mead, *op. cit.*, p. 26.

an administration building, some 75 cottages, a swimming pool, and a playground. A section of the city was set aside for the contractor's construction camp in which Six Companies was to build an office building, dormitories, a mess hall, cottages for married employees, a clubhouse, a hospital, a warehouse, a machine shop, a steam laundry, etc.[93] The contractor was given the right to use the water supply, sewage, and electric light and power systems of the city, but was required to pay a certain rental for these services and the land occupied. A transmission line was built from the Nevada-California Power Company's substation at the dam site to the city to provide power for the city's pumping plant and lighting system and to carry enough energy for all cooking, refrigeration, and other domestic uses.[94] The administration of the city government was placed in the hands of Mr. Walker R. Young, construction engineer for the Bureau of Reclamation; and Mr. Sims Ely was appointed city manager, under Mr. Young's supervision, to have direct charge of the city's affairs. United States deputy marshals were appointed as police and fire officers, and the duties of police judge were performed by a United States commissioner. A committee composed of representatives from the Bureau of Reclamation and from Six Companies was appointed to act in an advisory capacity to Mr. Young and Mr. Ely in all matters relating to the administration of Boulder City.[95]

One of the most difficult problems to be solved in the development of the city was the provision of an adequate water supply for the inhabitants and for the city's needs. There were only two sources of water available—the Colorado River and the artesian wells in the basin near Las Vegas. The water from the Las Vegas basin was clear and free from bacteria, but the construction of a pressure line 25 miles long, requiring a pump lift of 600 feet, would be necessary to bring the water to Boulder City. In addition, the adequacy of the supply was not definitely known, and the water was harder than that of the river. The Colorado River was only six or seven miles from the city; but

[93] *New Reclamation Era,* Vol. XXII, No. 6 (June 1931), p. 131.

[94] *Ibid.,* No. 2 (February 1931), p. 28.

[95] Wilbur and Mead, *op. cit.,* p. 28.

the pump lift necessary to bring the water from the low-water level to the top storage tank was over 2,000 feet. Further disadvantages of river water were its high silt content and its large bacteria count. In spite of these objections it was decided to use the river water because the supply was sufficient and determinable and because the cost of the system would be about $200,000 less than that for the artesian supply. To render the water acceptable for domestic use, however, it would be necessary to soften the water, to treat it chemically to destroy harmful bacteria, and to remove the silt content. The system adopted required the construction of four pumping stations which would lift the water to a receiving tank and filtration plant in Boulder City and force it through the distribution lines. At the filtration plant the water would be submitted to complete clarifying, chemical, and filtering processes, and would emerge at the end of the treatment perfectly clear, entirely free from harmful bacteria, and with hardness reduced to about 100 parts per million. The intake pumping plant was to be located about 3,000 feet downstream from the dam site, but it was planned to move the intake plant to a point above the dam when that structure was completed in order to procure clear water and to take advantage of the higher water surface to eliminate a part of the pump lift.[96] A complete distribution system was designed to cover the entire Boulder City area, and by 1933 nearly thirteen miles of mains and fifteen miles of service lines had been laid. At that time the water pumped from the river amounted to about 15,860,000 gallons per month and required 200,400 kilowatt-hours of electric power for pumping purposes. The cost of the entire system, including pumping, filtration, chemical treatment, and distribution, totaled approximately $470,000.[97]

The Bureau of Reclamation realized the importance of adequate health facilities, and included an extensive sewer system covering the area to be occupied and an efficient sewage-disposal plant as a part of the plan to make Boulder City a healthful

[96] This move was made, and the intake plant is now located above the dam. The city is not required to pay a storage fee for taking the water from Lake Mead. (Letter from the Bureau of Reclamation, Washington, D.C., December 1, 1938.)

[97] Wilbur and Mead, *op. cit.*, pp. 28–31.

place in which to live. The lines of the sewer system were laid in conjunction with the water distribution system, and preceded the construction of walks, curbs, and street paving. The sewage-disposal plant is of the separate sludge-digestion type and is designed to afford primary treatment for an average flow of 1,500,000 gallons a day. The cost of the sewage system was about $62,000 and that of the sewage-disposal plant about $25,000, or approximately $87,000 for the entire system. A health and sanitation board was appointed by the construction engineer to maintain general supervision of all sanitation features and to make frequent inspections of the sanitary measures in force in Boulder City.[98]

Of the initial appropriation for the Boulder Canyon Project, $525,000 was allocated to begin the laying out of Boulder City, the paving of the streets, the building of the water works, sewage system, and other conveniences, and the construction of a government office building and 25 houses for permanent government employees.[99] Before the end of June 1930 a headquarters office had been established in Las Vegas under the supervision of Mr. Young, the construction engineer, and surveys of the town site were started immediately. During the following year hundreds of inquiries were received by the Bureau of Reclamation from people who wished to establish various businesses in the new community. Since the business possibilities were limited, permits could not be granted to all the applicants without disastrous results, and a system was developed to limit the number of people who would be allowed to engage in business in the city. It was necessary to attract all of the types of business needed to serve the city satisfactorily, and although competition was considered desirable, an attempt was made to prevent it from becoming ruinous.[100] A circular of information was printed which outlined the rules for applications, and a schedule of rents was published for business and residence lots which varied in accordance with the area and desirability of the location. All applications for permits for the same line of business

[98] Wilbur and Mead, *op. cit.*, pp. 31–32.

[99] *New Reclamation Era*, Vol. XXI, No. 8 (August 1930), p. 146.

[100] *The Reclamation Era*, Vol. XXIII, No. 8 (August 1932), p. 137.

were carefully graded, and those applicants considered to be best fitted to serve the town were granted the permits. It was believed that in the future, when the town had reached its normal size, the extent of business opportunities would be more readily discernible and the necessity for limiting competition would largely disappear.

By the end of June 1931 the construction of the water system was well under way, and twelve of the government cottages were nearing completion. The government buildings were designed to become a permanent part of the city and were to be provided with heating and air-conditioning systems. On March 1, 1932, all offices of the project were moved from Las Vegas to Boulder City. The contractors' buildings were not as substantially built, and were constructed much more rapidly. By November 1931 Six Companies had completed a 1,300-man mess hall, a steam laundry, a machine shop, a warehouse, a garage, a clubhouse, a company store, six dormitories, and one hundred and twenty-two cottages, and had started construction on the hospital and the executives' lodge. The dormitories were all equipped with showers, lavatories, and combination heating and cooling systems, and the cottages contained all modern conveniences for comfortable living.[101] Before the end of the year the population had grown past the 2,500 mark, and a new city was thriving in the heart of the desert. It was estimated that the government had spent about $1,135,000 on Boulder City and Six Companies about $780,000, making a total of $1,915,880.[102]

Six Companies had completed its principal building program by April 1932, including 539 residences, four dormitories, and two office buildings, in addition to the buildings finished by November 1931. The Babcock and Wilcox Company planned to build a 79-room dormitory and a number of residences to house their employees who would soon begin work in the huge steel plant. Government permittees had erected buildings housing some fifty different types of business enterprises. These included a moving-picture theater, restaurants, drugstores, garages, service stations, tourists' camps, men's

[101] *New Reclamation Era*, Vol. XXII, No. 11 (November 1931), p. 234.

[102] *The Reclamation Era*, Vol. XXIII, No. 2 (February 1932), p. 32.

clothing stores, a dry-cleaning shop, a building-supplies yard, a lodging house, a hotel, a telegraph office, barber shops, beauty shops, a laundry, a mortuary, an electrical appliances store, and general stores. Churches were built by the Catholic, Episcopal, and Latter Day Saints denominations, and a community church also was established. An American Legion post was founded and a hall for it was erected. By the end of 1932 the Six Companies hospital, with room for 60 beds and adequately equipped with modern medical facilities, was finished, and the post office, employing a force of eight persons, had been moved into the new municipal building.[103] A survey in 1932 had shown that 851 children were residents of Boulder City, 651 of whom were of school age; and educational facilities became a necessary part of the building program. An eight-room school was built, and was maintained chiefly by Six Companies, since the children of its employees formed a large majority of the student body.[104] By 1933 the city had about 1,000 buildings and 15 miles of paved streets; and in 1934, during the peak of project construction, its population had grown to about 6,000, making it the third largest city in Nevada.[105]

After 1934, with the decline of construction activity in Black Canyon, the population of Boulder City gradually dwindled until in 1936 about 1,000 people remained as residents of the city.[106] The Six Companies contract had been completed, and all of the buildings which had formed a part of the contractor's construction camp were torn down. Many of the business establishments had closed their shops and had moved away. Some work remained to be done on the power plant, and when that was completed the city's population would be diminished still further. Boulder City was no longer a construction camp. It had been built to provide comfortable living conditions for thousands of men who were working on a great project under trying conditions; but that job was done. It has since contracted

[103] Wilbur and Mead, *op. cit.*, pp. 27–28.

[104] *Thirty-first Annual Report of the Commissioner of Reclamation for the Fiscal Year Ended June 30, 1932*, pp. 3–4, 28–29.

[105] *Engineering News-Record*, Vol. CXV, No. 26 (December 26, 1935), p. 878.

[106] Department of the Interior, Bureau of Reclamation, *General Information Concerning the Boulder Canyon Project* (1936), p. 4.

to a size only large enough to house the permanent staff employed in operating the dam and powerhouse and to accommodate the tourists who come to see the massive structures in Black Canyon and the lake which stretches back into the former valley of the Colorado. It is a permanent community, however, and may experience some revival in building activity in the near future if Lake Mead is developed successfully as a pleasure resort.

COMPARISON WITH OTHER PROJECTS

The Boulder Canyon Project is so much larger than any previous project of a similar type that it is difficult to picture its true size and importance without a comparison with other dams, reservoirs, and power plants which formerly had been considered to be extremely large structures. Recently other projects have been planned and placed under construction which will surpass the Boulder Canyon Project in several respects. Yet in spite of these new developments the Boulder Canyon structures will still hold their own as one of the most spectacular feats of engineering in the history of the world. The great dam with a height of 727 feet above bedrock is the highest dam ever built. It is 310 feet higher than the Owyhee Dam in eastern Oregon, its nearest competitor at present; and it is more than twice as high as the 354-foot Arrowrock Dam near Boise, Idaho, or the 328-foot Shoshone Dam in Wyoming, each of which at one time held the distinction of being the highest dam in the world.[107] The highest dam outside of the United States is the 414-foot Sautet Drac River Dam, a power dam in the French Alps.[108] Even the great Shasta Dam of the Central Valley Project in California and the Grand Coulee Dam of the Columbia Basin Project in Washington, both of which are now under construction, will not surpass the Black Canyon structure in height; the Shasta Dam will rise 560 feet above bedrock, while Grand Coulee will become the third highest dam with a height of 550 feet.

[107] Department of the Interior, Bureau of Reclamation, *Dams and Control Works* (1938), p. 1.

[108] *The World Almanac and Book of Facts,* 1938, p. 777.

The Boulder Canyon reservoir, now known as Lake Mead, is the greatest artificial reservoir ever constructed, and its capacity is far greater than that of any previous reservoir or of any reservoir now under construction. It is true that the 150-mile reservoir which will be formed behind Grand Coulee Dam will be much longer than Lake Mead, which stretches 115 miles into the valley above Black Canyon; but the 30,500,000 acre-feet capacity of Lake Mead is more than triple the capacity of the Grand Coulee reservoir with its 9,640,000 acre-feet. The Fort Peck Dam reservoir in Montana will have a capacity of 20,000,000 acre-feet when completed, which is nearly two-thirds that of Lake Mead, and will be the only reservoir which can reasonably be considered to approach the capacity of the Boulder Canyon reservoir. The greatest reservoir existing before the development in Black Canyon was the Assuan Dam reservoir in Egypt, with the comparatively small capacity of 4,060,000 acre-feet. The Boulder Canyon reservoir is large enough to hold about two years' average flow of the Colorado River, which is equal to approximately 10,000,000,000 gallons. When full it will hold enough water to cover all of the state of New York to a depth of one foot, and could provide a nine-year supply of water sufficient for all of the domestic needs of all of the inhabitants of the United States.[109] Topographical conditions in the vicinity of Boulder Canyon were ideal for the construction of a huge reservoir, and although higher dams may be built in future projects it is likely that Lake Mead will long remain the greatest artificial reservoir in the world.

In spite of its enormous height and the tremendous size of its reservoir, Hoover (Boulder) Dam will soon have serious rivals for the distinction of being the greatest dam in the world. Both the Grand Coulee Dam and the Shasta Dam will have a greater crest length and a greater volume of concrete than Hoover Dam, although neither one will be as high. Grand Coulee will have a crest length of 4,200 feet and a volume of 10,500,000 cubic yards, as compared with the Hoover's crest length of 1,282 feet and volume of 3,250,335 cubic yards. Shasta Dam will be

[109] Boulder Dam Service Bureau, *Boulder Dam Book of Comparisons* (1937), p. 14.

TABLE XII.—HOOVER (BOULDER) DAM COMPARED WITH OTHER GREAT DAMS*

Dam	Location	Cost	Year Completed	Maximum Height (feet)	Crest Length (feet)	Volume (cubic yards)	Reservoir Capacity (acre-feet)[a]
Hoover (Boulder)	Arizona-Nevada	$ 70,600,000	1936	727	1,282	3,250,335	30,500,000
Shasta	California	....	....	560	3,400	5,610,000	4,500,000
Grand Coulee	Washington	118,600,000	....	550	4,200	10,500,000	9,640,000
Hetch Hetchy	California	10,000,000	1938	427	....	....	359,000
Owyhee	Oregon	5,672,000	1932	417	833	556,471	1,120,000
Sautet Drac	French Alps	....	1936	414	....	....	106,000
San Gabriel	California	15,746,251	....	381	....	....	....
Arrowrock	Idaho	4,928,000	1915	354	1,100	602,200	286,500
Shoshone	Wyoming	1,439,000	1910	328	200	78,576	445,700
Esla	Zamora, Spain	12,000,000	....	328	....	....	810,800
Parker	Arizona-California	8,805,000	....	322	800	268,000	720,000
Elephant Butte	New Mexico	4,538,000	1916	306	1,162	605,200	2,637,700
Horse Mesa	Arizona	2,873,000	1927	305	784	147,357	244,900
Jandula	Andujar, Spain	5,000,000	1930	295	....	....	364,900
Friant	California	15,000,000	....	290	3,330	1,600,000	400,000
Roosevelt	Arizona	3,806,000	1911	284	1,125	242,970	1,420,000
Bartlett	Arizona	4,472,000	....	270	950	162,825	200,000
Norris	Tennessee	13,800,000	1936	265	....	....	3,401,500
Marshall Ford	Texas	17,700,000	....	265	2,325	....	3,068,900
Seminoe	Wyoming	4,360,000	....	265	560	161,000	1,000,000
Barberine	Switzerland	20,000,000	1921	259	....	....	254,100
Coolidge	Arizona	4,500,000	1928	249	....	....	1,200,000
Fort Peck	Montana	86,000,000	....	242	20,000	100,000,000[b]	20,000,000
Tygart River	West Virginia	15,700,000	....	232	....	....	327,100
Dnieprostroy	Russia	110,000,000	1932	200	....	....	895,500
Hartebeestpoort	South Africa	8,000,000	1923	193	....	....	1,258,200
Lloyd Barrage	British India	73,730,000	1928–32	190	....	....	....
Marathon	Greece	2,200,000	1929	177	....	....	33,100
Bonneville	Oregon-Washington	51,000,000	1938	170	1,090	1,000,000	480,000
Assuan	Egypt	29,000,000	1912	144	....	....	4,060,000
Sennar	Sudan	43,000,000	1928	128	....	....	429,700
Krishnaraja	British India	13,000,000	....	124	....	....	138,100
Imperial	Arizona-California	7,551,800	....	45	3,430	2,191,800	85,000
Laguna	Arizona-California	1,921,492	1909	40	4,844	451,000	....

* Bureau of Reclamation, *Dams and Control Works* (1938), pp. 1, 7, 20, 28, 31, 44–46, 49, 57, 85–86, 137, 258–60. See also Fred Locksley and Marshall Dana, *More Power to You* (1934), inside front cover, pp. 40, 68; *The Reclamation Era*, Vol. XXVIII, No. 5 (May 1938), p. 81; *Engineering News-Record*, Vol. CXX, No. 15 (April 14, 1938), p. 525; Bureau of Reclamation, *Grand Coulee Dam*; *World Almanac, 1938*, pp. 777–78; United States War [illegible] *Annual Report, 1936*, p. 517; United States War Department, Corps of Engineers, *Improvement of Columbia River at Bonneville, Oregon*

second only to Grand Coulee, with a crest length of 3,400 feet and a volume of 5,610,000 cubic yards.[110] A number of other dams, such as the Friant Dam of the Central Valley Project in California, the Imperial Dam, and the Laguna Dam, have much greater crest lengths than Hoover Dam but are smaller in volume and are still not as impressive as the structure in Black Canyon. At the time the Boulder Canyon Project was begun the amount of money to be spent in construction was considered a tremendous sum for a flood-control, water-storage, and power project. It was pointed out that the amount was as great as the combined cost of all previous reclamation projects in the Western states.[111] However, the cost of the Grand Coulee Project will be far greater than the cost of the Boulder Canyon Project,[112] and the dubious honor of being the most expensive reclamation development in the United States will pass from the Colorado River Basin to the Columbia.

The Boulder Canyon power plant will also be surpassed in size by the Grand Coulee power plant; but until the Grand Coulee structure is ready for operation the Boulder plant will be the greatest in the world. With an installed capacity of 1,835,000 horsepower, the Boulder power plant is far larger than Niagara (United States) with its 452,500 installed horsepower, than Conowingo with its ultimate capacity of 594,000 horsepower, than Muscle Shoals (Wilson Dam) with its ultimate capacity of 600,000 horsepower, than Bonneville with its ultimate capacity of 600,000 horsepower, and than the Dnieprostroy plant in Russia with its 750,000 horsepower.[113] The Boulder power plant will maintain its supremacy over the

[110] Actually the Fort Peck Dam in Montana will be far greater than either the Grand Coulee or the Shasta dams, with a crest length of 20,000 feet and a volume of 100,000,000 cubic yards. However, its height will be only 242 feet; and since it will be the hydraulic earth-fill type, it is not comparable with massive concrete dams such as Hoover, Grand Coulee, and Shasta. See Table XII.

[111] House Hearings before the Committee on Irrigation and Reclamation, 70th Congress, 1st Session, *Protection and Development of the Lower Colorado River* (1928), statement of Mr. H. S. McClusky, pp. 128–29.

[112] *The Reclamation Era,* Vol. XXV, No. 1 (January 1935), p. 14.

[113] *New Reclamation Era,* Vol. XXI, No. 6 (June 1930), p. 114; *The Reclamation Era,* Vol. XXVII, No. 7 (July 1937), p. 147. See also Fred Locksley and Marshall N. Dana, *More Power to You* (1934), p. 40.

Shasta Dam which will have an annual firm-power output of 1,500,000,000 kilowatt-hours as opposed to Boulder's 4,330,-000,000;[114] but the Grand Coulee power plant with an installed capacity of 2,700,000 horsepower and an annual firm-power production of 8,320,000,000 kilowatt-hours will become the undisputed leader of them all. The great turbines at the Boulder power plant, which are rated at 115,000 horsepower each, will be exceeded by Grand Coulee's turbines rated at 150,000 horsepower; and Boulder's generators, still the largest in the world at 82,500 kilovolt-amperes, will be far surpassed by the generators at Grand Coulee with their rating of 120,000.[115]

Thus within the next few years the Boulder power plant will take second place in the volume of power production, and aside from its height and the capacity of its reservoir Hoover Dam will no longer be the largest dam in the world. However, the importance of a project cannot be measured by its size alone, although the size of the great dam in Black Canyon is still sufficiently unusual to make it an object of world attention. The work of flood control, water storage, and river regulation is of inestimable value to the people of the lower Colorado River Basin, and the development will always stand out as one of the world's greatest flood-control and reclamation projects. Hoover Dam was the first of the gigantic dams to be built by the Reclamation Bureau, and as such it has blazed the trail for other projects, many of which will use the construction processes developed and the engineering knowledge gained during the building of the Boulder Canyon Project.

CONSTRUCTION OF THE ALL-AMERICAN CANAL AND THE IMPERIAL DAM

THE ALL-AMERICAN CANAL

The construction in Black Canyon was the largest and most spectacular part of the Boulder Canyon Project; but it was not by any means the only important feature of the development.

[114] *The Reclamation Era*, Vol. XXVIII, No. 5 (May 1938), p. 81.

[115] Department of the Interior, Bureau of Reclamation, *Grand Coulee Dam* (1938).

The project act specifically provided for the construction of a main canal located entirely within the United States, connecting the Laguna Dam or some other suitable diversion dam with the Imperial and Coachella valleys. The building of this canal and the Imperial Dam, which is to be used as a diversion point instead of Laguna Dam, formed an important part of the project's construction program and involved engineering problems which rivaled those of the Black Canyon structures in difficulty. As originally planned, the canal was to head at Laguna Dam and follow the main canal of the Yuma project to Siphon Drop, where it would take a southwesterly direction to Pilot Knob. From there it would follow a westward course through the sand-hill area to Imperial Valley. Opponents of the project had argued that the whole idea was impractical since it would be impossible to build a satisfactory canal through the sand-hill area, and that if such a canal should be constructed it would cost several times as much as the $38,500,000 allocated to this phase of the project. It was said that no permanent canal could be built in a district where even the hills were undergoing constant change and that the cost of keeping the canal from becoming clogged with silt would be prohibitive. These objections were not taken seriously, however, since before the Boulder Canyon Project Act had received the favorable action of Congress surveys had shown that the canal could be built satisfactorily within the cost estimates.[116] Engineers had no doubt of their ability to construct the canal, and with the improvements which had recently been made in mechanical equipment for excavating, handling material, etc., it was believed that the canal might cost considerably less than the original estimates. Undoubtedly the section of the canal in the sand-hill area would cause the greatest difficulty, but this was a problem of maintenance rather than of construction—it was known that some blow and drift sand would enter the canal, but observations had indicated that the amount would be surprisingly small. Winds of a transporting velocity blow for only some sixty days of the year in this part of the country, and if the banks of the canal

[116] House Report No. 918, 70th Congress, 1st Session, *Boulder Canyon Project* (1928), pp. 18–20.

were oiled or lined with vegetation the amount of sand that would enter the canal could easily be carried in suspension by the water. It was suggested that the sand-hill section of the canal should be lined with concrete and that a suitable gradient should be provided to produce a velocity of flow high enough to carry large amounts of sand, but this was not considered essential. In any event, there seemed to be no ground for counting this problem as one of controlling importance, and it was unlikely that the maintenance expense involved would ever be great enough to place a serious burden upon the operation of the canal.[117]

The surveys and investigations necessary to determine definitely the location and estimated cost of the All-American Canal were begun in 1929. On May 23, 1930, all survey records were taken to the Denver office of the Bureau of Reclamation, where the designs and estimates were completed. It was discoverd that if the canal diverted water from the river at a point about five miles above Laguna Dam (the original diversion point contemplated), 22 feet would be gained in elevation, resulting in a large saving of excavation in the long deep cut through the sand-hill region west of Pilot Knob. There would also be an increase in the power possibilities on the canal and an addition of about 26,000 acres of land to the area which could be irrigated by gravity. For these reasons it was decided to build the Imperial Dam and the diversion works at the point some five miles above Laguna Dam. The canal was designed to begin with a capacity of 15,000 second-feet and to become smaller at definite locations according to the provisions later inserted in the All-American Canal contract.[118] It was to have a maximum width of 200 feet at water surface and 134 feet at bottom, with a depth of 22 feet. It was to be 80 miles long, and the Coachella branch was to measure 130 miles in length.[119] Four

[117] Department of the Interior, Bureau of Reclamation, *Development of the Lower Colorado River* (1928), report of Professor W. F. Durand, special adviser to the Secretary of the Interior, pp. 389–91.

[118] Ray Lyman Wilbur and Northcutt Ely, *The Hoover Dam Power and Water Contracts and Related Data* (1933), pp. 326–27.

[119] It is interesting to note that the Friant-Kern Canal, one of the important features of the Central Valley Project in California, will be relatively small as com-

miles of the All-American Canal and 47 miles of the Coachella branch would require lining; and it would be necessary to build siphons or culverts under the numerous washes, which were ten in number on the main canal and ninety on the branch. The total excavation was estimated at 54,000,000 cubic yards. It was planned to develop power at Pilot Knob and at four other drops on the canal; and plans were also drawn, but not officially adopted, to extend the distance of the main canal 63 miles and to install a pumping plant to irrigate lands on the west mesa. It was estimated that 842,507 acres of land west of Pilot Knob would be subject to irrigation by gravity when the All-American Canal system was completed, and that 188,572 additional acres could be irrigated with pump lifts varying from 50 to 235 feet.[120]

In November 1933 the sum of $6,000,000 was appropriated by the Public Works Administration to begin the construction of the canal. Bids were received on June 7, 1934, and on June 23 a contract for $4,859,587 was awarded to the W. E. Callahan Company of St. Louis, Missouri, and the Gunther and Shirley Company of Dallas, Texas, for the excavation of about 31 miles of main canal through the sand-hill area.[121] Actual work was started on August 8, and by July 1, 1935, about one-third of this contract was completed.[122] During the year's work it was discovered that the sand hills were underlain with gravel and their menace was thus greatly reduced. The sand on the banks of the canal could easily be covered with a sheathing of this heavy gravel in order to prevent the banks from blowing into the canal on one side or from blowing away on the other.

From time to time other bids were received, and other contracts for excavation were awarded. By 1936 work was under way on most of the major features of the project. Up to March 19 of that year a total of 30,503,947 cubic yards of material

pared with the All-American Canal, in spite of the publicity concerning its great size. It will be 68 feet wide at water surface, 30 feet wide at bottom, and 15 feet deep in its upper reaches. Its initial capacity will be 3,500 second-feet, or less than one-fourth of that of the All-American Canal. See *The Reclamation Era*, Vol. XXVIII, No. 5 (May 1938), p. 81.

120 Wilbur and Mead, *op. cit.*, p. 71.

121 *Engineering News-Record*, Vol. CXII, No. 26 (June 28, 1934), p. 851.

122 *Annual Report of the Secretary of the Interior for the Fiscal Year Ended June 30, 1935*, pp. 54–55.

had been removed, which accounted for approximately half of the excavation required for the main canal.[123] It was decided to build the substructures of the powerhouses concurrently with the construction of the canal, since concurrent construction would involve a saving of about $500,000. There are six points along the canal where there will be sharp declines suitable for the development of power, but the development of two of these will be postponed until such time as there is need for them. According to the plan, the irrigation districts, and not the Bureau of Reclamation, will install the machinery and operate the power plants. Much of the power will be used to pump water to the higher lands which cannot be reached by gravity,[124] and its production was considered a local rather than a national affair.

The main line of the All-American Canal was about 85 per cent complete by the end of 1936, and by the end of 1937 most of the excavation work had been finished. Twelve construction contracts were under way during the year on building compacted lining, waste overchutes, turnouts, drainage inlets, wasteways, railroad and highway bridges, siphons, flumes, river crossings, and power drops. Four more contracts were let in 1938 to bring under construction all of the remaining structures on the main canal.[125] Excavation was completed during that year, all structures along the first 36 miles were finished, and water was diverted for the first time on October 18. In August excavation on the first 40 miles of the Coachella branch of the canal was begun, and specifications for the next 40 miles were in preparation.[126] Thus by the beginning of 1939 the All-American Canal phase of the Boulder Canyon Project was rapidly nearing completion.

IMPERIAL DAM

In November and December, 1931, investigations were made of the Imperial Dam site, and studies were carried on to determine the best design for the desilting works to be built in con-

[123] *Engineering News-Record,* Vol. CXVI, No. 18 (April 30, 1936), p. 629.
[124] *Ibid.,* Vol. CXVII, No. 22 (November 26, 1936), p. 765.
[125] *Ibid.,* Vol. CXX, No. 6 (February 10, 1938), p. 223.
[126] *Ibid.,* Vol. CXXII, No. 5 (February 2, 1939), p. j.

nection with the dam.[127] Since the All-American Canal will head at the new dam and will carry water to the Yuma Canal which formerly was supplied by water diverted at Laguna Dam, the old diversion works will become practically obsolete except as a possible tail-water control for the new dam. Imperial Dam will be the slab-and-buttress type with a total length of 3,430 feet. The maximum height will be 45 feet, and the water surface will be raised 23 feet. A reservoir with a capacity of 85,000 acre-feet and a surface area of 7,500 acres will be created, although this storage space is not required by the project. Actually the reservoir will soon become filled with silt in spite of the fact that the Boulder Canyon development and Parker Dam will control all but 10,000 square miles of the drainage area of the river down to that point. It has been estimated that with a diversion of 12,000 second-feet the silt load entering the canals would be some 60,000 tons per day. The silt would soon obstruct the flow and require excavation if desilting works were not built. Elaborate works for this purpose are under construction, and it has been computed that their cost of about $1,500,000 will save a yearly expenditure of some $1,000,000 for excavation from the canals. The cost of Imperial Dam plus the desilting works will amount to about $7,551,800.[128]

Designs and specifications for the Imperial Dam were in the course of preparation during the first half of 1935, and bids were called for late in the year. The contract for the construction of the dam and the desilting works was let in November 1935 to the Morrison-Knudsen Company of Boise, Idaho, the Utah Construction Company of Ogden, Utah, and Winston Brothers Company of Minneapolis, Minnesota, at their joint bid of $4,374,240 (not including the cost of materials to be furnished by the government). Work was started in January 1936,[129] and by January 1937 a pile-trestle bridge had been

[127] *Thirty-first Annual Report of the Commissioner of Reclamation for the Fiscal Year Ended June 30, 1932*, p. 31.

[128] Department of the Interior, Bureau of Reclamation, *Dams and Control Works* (1938), pp. 137–42.

[129] *Annual Report of the Secretary of the Interior for the Fiscal Year Ended June 30, 1936*, p. 62.

built across the river. Cofferdams were built, and during the following summer the river was diverted around the dam site. Since storage had been begun in Lake Mead, only relatively short flood periods were expected, principally from the Williams River, and no construction difficulties were anticipated. Imperial Dam is being built primarily as a diversion dam. It was not designed to do flood-control or water-storage work and is of little value in rendering those services. It is, however, an indispensable part of the All-American and Gila canal systems, and as such it holds an important place in the development of the lower Colorado River Basin.

CONSTRUCTION OF THE COLORADO RIVER AQUEDUCT AND THE PARKER DAM

THE COLORADO RIVER AQUEDUCT

Officially the Colorado River Aqueduct is not a part of the Boulder Canyon Project. It is an independent project, separately financed, and controlled by the Metropolitan Water District of Southern California. Yet it is so closely connected with the development in Black Canyon and its affairs are so interrelated with those of the dam and the power plant that it is impossible to give a complete picture of one project without some discussion of the other. The Boulder Canyon Project Act did not mention specifically the Colorado River Aqueduct; but the provisions for the sale of stored water contemplated the sale of some of that water to certain cities in southern California and thus the construction of an aqueduct from the river to the Pacific coast. The possibilities of such a project had long been a topic of discussion in southern California; but the cities composing the Metropolitan Water District had realized that it would be useless to build an aqueduct to the Colorado River unless the river were regulated and a steady supply assured. They had also realized that plenty of cheap power must be available to pump the water through the aqueduct over the mountains to the coastal plain. The construction of the Boulder Canyon Project fulfilled these requirements and made the operation of the aqueduct a possible and reasonably economic undertaking. A number of the larger cities in southern California had been in great need of additions to their water supplies, and had looked to the Colorado River as the logical source of such additions. As a result they gave their active support to the Boulder Canyon Project and helped to secure Congressional action favorable to the project legislation. Later the Metropolitan Water District entered into contracts with the United States to buy large amounts of Boulder power and water and to pay for the building of the Parker Dam.[130] Thus the Colorado River Aqueduct will create an additional, important market for Boulder Canyon products and will add to the safety of the project's financial plan. The Parker Dam not only will serve as a necessary feature of the aqueduct system but will provide still further facilities for the development of the lower Colorado River Basin.

For more than thirty years the City of Los Angeles has faced the problem of finding an adequate supply of water for her growing needs. By 1906 it had become obvious that the Los Angeles River was no longer a dependable source of supply, and the construction of the Los Angeles Aqueduct to bring water from the Sierra-Nevada Mountains, 250 miles away, was begun. At that time the population of the city was approximately 160,000; but with the completion of the first aqueduct in 1913 at a total cost of $24,000,000 it was estimated that there was sufficient water to supply the needs for a population of 2,000,000.[131] It was thought that the water supply problem had been solved for many years to come, but the Los Angeles metropolitan area grew so rapidly that the search for an additional supply of water began soon after 1920. In 1923 Mr. William Mulholland, the builder of the Los Angeles Aqueduct, pointed out that the Colorado River was the only source of sufficient magnitude to meet the future needs of the city.[132] Surveys were begun, and filings were made for the appropriation of 1,500

[130] Wilbur and Ely, *op. cit.*, pp. 326–27.

[131] E. F. Scattergood, "Engineering and Economic Features of the Boulder Dam," *The Annals of the American Academy of Political and Social Science*, Vol. CXXXV, No. 224 (January 1928), pp. 117–18.

[132] Don J. Kinsey, *The River of Destiny* (1928), pp. 50–51.

second-feet, or about 1,000,000 acre-feet per annum for domestic purposes.[133] Before 1928 various possible routes for the aqueduct had been charted, some 50,000 square miles of rugged mountain and desert country between the river and the coastal plain had been surveyed, the citizens of Los Angeles had voted two bond issues to finance the surveys and the preliminary construction work, nearly $2,000,000 had been spent for engineering right of ways,[134] and the Metropolitan Water District was in the process of formation. After the passage of the Boulder Canyon Project Act it became necessary to select a definite route for the aqueduct and to draw up the engineering plans. This was no easy task, because approximately sixty routes had been suggested. All of the so-called "gravity" routes were eliminated first, since they would require diversion far up the river in order to gain sufficient altitude and would necessitate the construction of extremely long aqueducts. A gravity aqueduct would involve such a high construction cost that the savings to be gained through the elimination of future pumping costs would probably not be worth the additional investment required.[135] Months were then spent in examination of the various "pump" routes; and finally, in December 1930, the Parker route was officially approved by the board of directors of the Metropolitan Water District.[136]

The Parker route for the aqueduct was planned to start at the Parker Dam Reservoir a few miles north of Parker, Arizona, and about 155 miles south of Black Canyon. The water would flow to an intake pumping station and be pumped by this and four other pumping stations over the mountains to the Cajalco Reservoir, 10 miles south of Riverside. A 237-mile power line was to be built from the Boulder power plant to the five pumping stations, although the system would develop some

[133] *Engineering News-Record,* Vol. XCIII, No. 7 (August 14, 1924), p. 277, and Vol. XCIV, No. 14 (April 2, 1925), p. 560.

[134] *Ibid.,* Vol. XCIX, No. 9 (September 1, 1927), p. 360.

[135] Senate Hearings before the Committee on Irrigation and Reclamation, 69th Congress, 1st Session, *Colorado River Basin* (1925), Part 1, statement of H. A. Van Norman, representing the City of Los Angeles, p. 120.

[136] *Engineering News-Record,* Vol. CV, No. 22 (November 27, 1930), p. 854, and No. 26 (December 25, 1930), p. 1024.

power of its own. The maximum lift above the Parker Reservoir would be 1,617 feet, and the length of the aqueduct from the Colorado River to the terminal reservoir would be 242 miles. This route was selected above all others because construction hazards were the smallest and safety against earthquake damage the greatest, and because it required the construction of no unusually large tunnels. The route would be less expensive in first cost than any of the others, and would have a smaller operating cost because of a lower pump lift.[137] According to the plans the aqueduct would comprise 29 tunnels 16 feet in diameter with a total length of 92.1 miles, 53 sections of covered concrete conduit with a total length of 54.5 miles, 98 sections of open concrete-lined canal with a total length of 62.8 miles, 146 inverted siphons of various designs with a total length of 28.7 miles, and several reservoirs, pumping stations, and appurtenances with a total length of 3.6 miles.[138] The Cajalco terminal reservoir would have an immediate capacity of 100,000 acre-feet, and an ultimate capacity of 225,000 acre-feet. This reservoir would be the intake for the distribution system to the cities of the Metropolitan Water District, which in 1938 included Anaheim, Beverly Hills, Burbank, Compton, Fullerton, Glendale, Long Beach, Los Angeles, Pasadena, San Marino, Santa Ana, Santa Monica, and Torrance. The distribution system would comprise nine tunnels of 16 miles total length and steel-pipe lines of 156 miles total length, making an aggregate distribution main length of 172 miles of varying diameters. The whole distribution system would be gravity flow made possible by the 1,405-foot elevation of the Cajalco Reservoir. From Cajalco to the north and then to the west for a total distance of 45 miles the main line would run to the Norris Reservoir in San Gabriel Canyon. This would be a secondary storage point from which distribution mains would continue to the west and south.[139] The entire aqueduct project, including the distribution system, would require about 6,800,000 cubic

[137] *New Reclamation Era*, Vol. XXII, No. 2 (February 1931), p. 37.

[138] Department of the Interior, Bureau of Reclamation, *Boulder Canyon Project—Questions and Answers* (1936), pp. 16–17.

[139] *Engineering News-Record*, Vol. CXVIII, No. 19 (May 20, 1937), p. 747.

yards of embankment, 45,000,000 cubic yards of excavation, 5,000,000 cubic yards of concrete, and 7,500,000 barrels of cement. The cost of the aqueduct was originally estimated at $200,000,000; but with the ultimate installation of the entire system, the cost will probably rise to over $280,000,000, an amount far greater than the $165,000,000 maximum to be appropriated for the Boulder Canyon Project.

On September 29, 1931, by a vote of nearly five to one, the citizens of the cities comprising the Metropolitan Water District authorized a bond issue of $220,000,000 to finance the building of the Colorado River Aqueduct. If the condition of the money market had been normal in 1931 and 1932, the district could probably have sold its bonds locally. With conditions as they were, however, the district was forced to appeal to the Reconstruction Finance Corporation, which was the only bidder for the first block of bonds offered.[140] These bonds were made payable in fifty years, and the first installment on principal was to become due fifteen years after the date of sale. All bonds sold to the R.F.C. bore interest at 5 per cent; but the Corporation agreed to make a 1 per cent refund to the district until April 1, 1939. In this manner $40,000,000 of the district's bonds were sold during 1933 and 1934. By April 1938, the R.F.C. had purchased $207,000,000 of the bonds, and $1,500,000 had been sold to the Public Works Administration. At that time it was estimated that the aqueduct from the Colorado River to the Cajalco Reservoir could be completed at a cost within the $220,000,000 bond issue voted in September 1931,[141] and that the original cost estimates were thus very close to the actual costs incurred during construction.

On June 19, 1932, Congress transferred to the Metropolitan Water District the public and reserved lands necessary for the right of way,[142] and on December 12 a contract amounting to about $7,331,815 was awarded to the Metropolitan Engineering Corporation for the construction of the 12.7-mile San Jacinto tunnel. A few days later Mr. Frank E. Weymouth, chief en-

[140] *Engineering News-Record,* Vol. CIX, No. 24 (December 15, 1932), p. 732.

[141] *Ibid.,* Vol. CXX, No. 13 (March 31, 1938), p. 483.

[142] *United States Statutes at Large,* Vol. XLVII, Part 1, pp. 324–26.

gineer and general manager of the Metropolitan Water District, authorized the beginning of work on 33 miles of tunnels through the Little San Bernardino Mountains northeast of the Coachella Valley. It was estimated that this latter work, which was begun on force account to avoid delay, would cost $18,000,000 and would require five and one-half years to complete.[143] By June 1933 nineteen contracts were under way on the project. Arizona's suit in 1935 to prevent the completion of Parker Dam did not delay work on the aqueduct, since the Metropolitan Water District had an alternative plan to pump the water directly from the river in case work on the dam should be suspended permanently, and the building of the aqueduct continued with no further legal difficulties to retard its progress.[144] In July 1937, when the 237-mile transmission line from the Boulder power plant was completed, about 64 per cent of the aqueduct project had been constructed, and in December work was begun on the Gene Wash Dam to create a pool which would serve as a reservoir for the first pump lift above Parker Dam and as a sump for the second lift. Later work was started on another reservoir at Copper Basin, and construction was begun on the five pumping plants at Intake (on the Parker Dam Reservoir), Gene Wash, Iron Mountain, Eagle Mountain, and Hayfield. The greatest difficulty encountered on the project was in the building of the San Jacinto tunnel, where unprecedented water inflows delayed the work time after time; but the tunnel was finally drilled through on November 19, 1938. By that time all other tunnels had been finished, and the aqueduct was 90 per cent complete.[145] Construction on the main line of the Colorado River Aqueduct was finished in 1939, and the first delivery of water into the Cajalco Reservoir was made in November of that year.[146] Most of 1940 was to be devoted to testing and filling the aqueduct reservoirs; but by 1941 the distribution of Colorado River water to the cities of the Metropolitan Water District would begin.

143 *Engineering News-Record*, Vol. CIX, No. 26 (December 29, 1932), p. 795.
144 *Ibid.*, Vol. CXIV, No. 19 (May 9, 1935), p. 687.
145 *Ibid.*, Vol. CXXI, No. 21 (November 24, 1938), p. 629.
146 *Ibid.*, Vol. CXXIV, No. 7 (February 15, 1940), p. 260.

PARKER DAM

The final plan for the diversion of water into the Colorado River Aqueduct contained a provision for the construction of a dam across the river to serve as a diversion point, although this feature was not contemplated in the original proposal. The Metropolitan Water District had planned to pump the water directly from the river;[147] but it was obvious that a reservoir at the diversion point would reduce the pump lift, insure a steady supply, and help to desilt the water. To gain these advantages, it was eventually decided to build a dam about fourteen miles north of Parker, Arizona, and just below the mouth of the Williams River. The primary purpose of the dam would be to serve as a diversion point for the aqueduct; but it would also control the occasional floods of the Williams River and would serve to provide water storage and power for the Colorado River Indian Reservation and the Parker-Gila project. According to the plans the dam was to be of the concrete, variable-radius-arch type with a crest length of 800 feet, a base width of 100 feet, a top width of 50 feet, a maximum height above bedrock of 322 feet, and a volume of about 268,000 cubic yards.[148] It was known that the detritus overlying sound rock at the Parker Dam site was very deep, and that upon completion nearly three-fourths of the dam would be below the bed of the river. It was estimated, therefore, that the river level would be raised only about 70 or 80 feet but that the reservoir, 28,000 acres in area and 720,000 acre-feet in capacity, would stretch 45 miles up the Colorado River and 5 miles up the Williams.[149] The intake for the aqueduct was to be about 8,000 feet above the dam on the California side. From 80,000 to 100,000 horsepower of electric energy were to be generated at the dam, half of which, according to the contract, was to be used by the federal government on irrigation projects. The dam

[147] *New Reclamation Era,* Vol. XXII, No. 2 (February 1931), p. 37.

[148] Department of the Interior, Bureau of Reclamation, *Dams and Control Works* (1938), p. 260.

[149] In 1939 the name "Havasu Lake" was adopted by the Department of the Interior's Board of Geographic Names as the official designation of the reservoir created by Parker Dam. See *Engineering News-Record,* Vol. CXXII, No. 23 (June 8, 1939), p. 747.

was to be built by the Reclamation Bureau; but the estimated cost of $13,000,000 was to be paid by the Metropolitan Water District. The contract between the United States and the district for the construction of the dam was signed on February 10, 1933.[150]

Construction of Parker Dam was expected to be undertaken during the filling of the Boulder Canyon reservoir, and the government advertised for bids for the work in 1934. The bids were opened on August 23, and on September 10 the contract was awarded to Six Companies, Inc., which had submitted the low bid of $4,239,834.[151] The Public Works Administration allotted $2,000,000 for the initial construction operations on this project, and work was begun in October 1934.[152] During the following month and again early in 1935 construction was halted by Arizona's efforts to block the project through the courts, and the work was not resumed until the end of August, when the legal difficulties were overcome. Diversion-tunnel excavation was completed in April 1936, and by August all of the concrete for lining had been poured in both tunnels. The river was diverted in October, aided by the construction of upstream and downstream cofferdams. Excavation of the foundation for the dam was started immediately, and bedrock was reached on July 21, 1937, some 240 feet below the bed of the river. That excavation was believed to be the deepest ever dug for the foundation of a dam.[153] Ten days later the first bucket of concrete was poured in the foundation, and by the end of the year most of the concrete was in place. On July 1, 1938, the diversion tunnels around the dam site were permanently sealed with concrete, and the river was returned to its normal channel to build up the reservoir and to flow over the completed crest of the dam.[154]

Parker Dam is the third dam to be built on the lower Colo-

[150] Wilbur and Ely, *op. cit.*, pp. 311–14.

[151] *Annual Report of the Secretary of the Interior for the Fiscal Year Ended June 30, 1935*, p. 47.

[152] *Engineering News-Record*, Vol. CXIII, No. 9 (August 30, 1934), p. 286, and No. 15 (October 11, 1934), p. 476.

[153] *Ibid.*, Vol. CXIX, No. 5 (July 29, 1937), p. 164.

[154] *Ibid.*, Vol. CXXII, No. 5 (February 2, 1939), p. B.

rado River since the completion of Laguna Dam in 1909. It is the second great dam to be built on the river for flood-control and water-storage as well as for diversion purposes, and aside from the fact that it is a part of the Colorado River Aqueduct system it will serve to reregulate the river 155 miles below Black Canyon. The importance of Parker Dam is dwarfed by comparisons with Hoover Dam; but it is truly a great step toward the complete control and utilization of the river. Before 1931 no attempt had been made to regulate the flow of the lower Colorado by damming it in one of its canyons; but now, less than eight years later, two great structures have been built to achieve that purpose. During the past few years rapid progress has been made in the building of reclamation projects in this section of the Southwest, and with the completion of construction in Black Canyon on the All-American Canal system and on Parker Dam another great step will have been taken toward the ultimate development of the lower Colorado River Basin. One of the important reasons for this progress is the rapid advance made recently in engineering knowledge, through which projects previously considered impossible are now within the limits of economic construction. With this knowledge, and with the first great projects nearing completion, it is reasonable to expect the building of other large structures on the Colorado to carry on the reclamation work so successfully begun with the Boulder Canyon development.

Chapter VI

ECONOMIC EVALUATION OF THE PROJECT'S SERVICES

The Boulder Canyon Project will in all probability prove to be a successful investment from the financial point of view. The cost of the project will undoubtedly be returned to the government, the states of Arizona and Nevada will receive a substantial share of the project's revenues, and a fund will be created for use in furthering the development of the Colorado River Basin.[1] However, the financial success or failure of a project cannot be taken as the sole measure of its value from an economic point of view. Questions such as whether or not there is a real need for the project's services and whether or not the particular project conforms to the best method of securing the services still remain unanswered. In the case of the Boulder Canyon Project the most important services rendered are those of flood control and river regulation, storage of water for agricultural and domestic uses, and production of electric power. Less important services include the employment of thousands of workers during a period of business depression, the creation of a new recreational area, and the founding of a sound basis upon which to proceed with the further development of the Colorado River Basin. It is the purpose of this chapter to analyze these various services as elements of future events which may react favorably or unfavorably to the public interest. The Boulder Canyon Project is a development of national importance; and the services not convertible directly into money, as well as those which are salable, must be considered in making this analysis. It is upon the broad basis of public interest, rather than upon the prospects of financial gain alone, that the economic justification of the project must rest.

[1] See Table XI, p. 184, above.

FLOOD CONTROL AND RIVER REGULATION

Since navigation is no longer an important feature in the economic development of the Colorado River, flood control for the protection of the agricultural areas of the lower basin is now the most valuable service performed by the Boulder Canyon Project through the regulation of the stream's flow. The need for flood control is undisputed. The disastrous floods which have occurred in the Imperial Valley, at Yuma, at Palo Verde, at Needles, and elsewhere on the lower river are ample proof of the urgent necessity for protection to agricultural lands from the unpredictable torrents of the stream. That these lands are worth saving is also undisputed. They are among the most productive lands in the United States; they are worth hundreds of millions of dollars; and they are now supporting thriving communities which are dependent upon the surrounding agricultural territory for their continued prosperity. If the Colorado River should ever change its course permanently from the Gulf of California to the Salton Basin, about 1,250,000 acres of excellent farm land in the Imperial and Coachella valleys would gradually be submerged, and the communities of Imperial, El Centro, Holtville, Calexico, Mecca, Thermal, Coachella, and even Indio would be covered with water. Ultimately the water would cover some 1,900 square miles of land in California and 100 square miles in Lower California without hope of an early restoration, since it would take approximately forty years to fill the basin and fifty years more to dry up the inland sea. The deepening channel of the river, caused by back-cutting, would destroy the Yuma project and perhaps other agricultural areas farther north. All of the delta lands would be robbed of their irrigating water and would quickly become desert wastes. Such a catastrophe would not only mean the ruin of the best improvements in the lower Colorado River Basin, but would also represent a loss which the nation as a whole could ill afford to suffer. Before the construction of the Boulder Canyon Project, the constant flood danger made capital available in the Salton Basin only at high rates of interest, thus retarding to a degree the development of this great national resource. Other projects on the river, although not subject to permanent inundation, also faced the flood menace; and in some sections even the Federal Farm Loan Banks refused to lend money because of this condition. The happiness of the settlers, the security of their property, and the proper development of this highly productive area thus depended largely upon adequate flood control.

Experience with previous floods had indicated that the best way to effect satisfactory flood control would be to modify the high peaks of flow and to prevent the deposition of the large quantities of silt that would choke the stream bed. The Boulder Canyon Project performs this service through the use of its huge reservoir, in which large quantities of silt are being stored and from which the stream may be regulated in such a way as to dispense with both the peaks and the depressions in rate of flow. According to the plan, the top 72 feet of the reservoir, or 9,500,000 acre-feet of the 30,500,000 estimated capacity of the reservoir, will be kept empty (or reserved) for flood control, and will be filled only during the flood stages of the upper river. About 7,000,000 acre-feet will be reserved for silt storage, and about 14,000,000 acre-feet will be reserved for water storage for agricultural and domestic uses.[2] It has been estimated that reservoir space of 8,000,000 to 10,000,000 acre-feet reserved for flood control is sufficient to regulate most of the floods of the river to 40,000 second-feet; that is, floods entering the reservoir at a rate as high as 200,000 second-feet can be curbed to 40,000 second-feet or less upon leaving the reservoir. Unusual floods of 300,000 second-feet or more, even when prolonged for some time, can be reduced to a 75,000 second-foot flow.[3] There is no danger from flood on the section of the river from Yuma to the Gulf until the river reaches a stage of 50,000 second-feet or more, and with the Boulder Canyon reservoir as a regulating agency this figure will be exceeded only occasionally for very short periods of time. Since the Parker Dam reservoir will act as a reregulating agency, the irrigated lands in the lower basin

[2] Department of the Interior, Bureau of Reclamation, *Boulder Canyon Project—Questions and Answers* (1936), p. 4.

[3] *Engineering News-Record*, Vol. CXIII, No 22 (November 29, 1934), p. 688. See also *The Reclamation Era*, Vol. XXVII, No. 2 (February 1937), p. 26.

no longer face a flood menace of serious proportions from the upper river, and the Boulder Canyon Project has thus solved this phase of the flood problem. In the future, when other reservoirs are built upstream, the space reserved for flood control in Lake Mead may be reduced safely to 5,000,000 acre-feet or less, thus releasing additional space for silt and water storage without endangering the effectiveness of the reservoir for flood control.[4]

The estimated silt load of the river at Lake Mead is 137,000 acre-feet per annum.[5] If the river continues to bring that amount of silt to the reservoir, the silt storage space of 7,000,000 acre-feet will be filled in about fifty-one years. With upstream developments, however, the average silt load will decrease. The construction of reservoirs on the tributaries above Boulder Canyon, especially the construction of power reservoirs on the lower parts of the main tributaries, will decrease materially the volume of silt reaching the lake. A power dam at Diamond Creek, for instance, would stop for many years a large part of the silt which otherwise would pass on to the Boulder Canyon reservoir. It is estimated, therefore, that in fifty years the total silt deposits will not exceed 3,000,000 acre-feet.[6] At this rate the silt storage space will not be fully occupied for well over one hundred years, and even after silt deposits have begun to encroach upon the space reserved for water storage the effective use of the reservoir for flood control will probably not be impaired for another three hundred years or more.[7] It should not be forgotten, however, that the silt will enter the reservoir very rapidly. It is only because the reservoir is so large that it will have such a long life. A small reservoir would soon lose its effectiveness for flood control and would thus not be worth while from an economic point of view.

The Boulder Canyon Project is not a complete solution to the silt problem either permanently or temporarily, but it is an

[4] Frank E. Weymouth, "Major Engineering Problems: Colorado River Development," *The Annals of the American Academy of Political and Social Science*, Vol. CXLVIII, No. 237 (March 1930), Part 2, p. 25.

[5] See p. 12, above.

[6] Department of the Interior, Bureau of Reclamation, *Boulder Canyon Project—Questions and Answers* (1936), p. 5.

[7] *Engineering News-Record*, Vol. CXIV, No. 11 (March 14, 1935), p. 399.

effective partial solution. Silt will still enter the lower Colorado from the Williams and Gila rivers, but the clear water which emerges from Lake Mead will have a greater capacity to pick up sediment from the river bed than the muddy stream which formerly flowed through Black Canyon. This will result in a gradual deepening of the bed, with increased security to all communities on the lower river.[8] Since the river grade is low, however, only the finer silt will be carried along, and the coarser will be left on the bed of the river. In this way the stream bed will eventually be paved with coarse material which it has no power to cut, and the river will become clear for many miles below the dam. The building of the Parker Dam below the mouth of the Williams River and future developments on the Gila and its tributaries will also help greatly to curtail the quantity of silt reaching the lower river.

Before the development in Black Canyon the instability of the river channel could be attributed partly to the practice of diverting water for irrigation. At the diversion point as much sediment as possible was taken from the water and was dumped back into the stream to be handled by the reduced flow. The ideal condition so far as flood protection is concerned is to regulate the flow so that it will always be uniform below the irrigating diversions. Because there are other uses for the water, this goal will never be reached; but the regulation of the river through the Boulder Canyon reservoir will help to approach the ideal and thus bring about a satisfactory solution to another aspect of the silt problem.[9]

The Boulder Canyon Project, therefore, presents an effective method of achieving flood control through the regulation of the stream's flow and through at least a partial solution of the silt problem. There is still the question, however, whether or not the project is the most economic method of securing the greatly needed flood-control services. Broadly, there are only two ways to secure flood control on the lower Colorado River. The first is to build levees and to dredge the stream bed in order

[8] House Report No. 2864, 71st Congress, 3d Session, *Protection of the Palo Verde Valley* (1931), p. 7.

[9] Walter H. Voskuil, *The Economics of Water Power Development* (1928), p. 95.

to keep the river in its channel; and the second is to build a reservoir in one of the canyons of the river to hold back the floods, as in the case of the Boulder Canyon Project, although perhaps not on such a gigantic scale.

The history of the floods on the lower river, especially in the Imperial Valley, has shown the futility of the first method; but some of the opponents of the Boulder Canyon Project still urged the building of levees as a more economic method of protecting the agricultural areas of the lower river. It was stated that flood control could be secured through channel dredging and levee construction at a cost not to exceed $6,000,000[10] and that it was ridiculous to spend the $70,000,000 or more required by the Boulder Canyon Project for the construction of the dam and reservoir. In the writer's opinion this statement was not based upon sound reasoning or accurate information. In the first place, the entire cost of the dam could not be attributed solely to flood control, since water storage and power production were other services which should bear a large share of the cost. In the second place, far more than $6,000,000 had already been spent on levee construction with very doubtful success, and there was no proof that further expenditures would insure flood control. No construction short of a complete line of embankments adequately maintained on both sides of the river from Laguna Dam to the Gulf could be considered as an adequate solution to the problem, and such construction would involve an insupportable expense.[11] Past experience had shown that the levees, no matter how well maintained, could not be depended upon during periods of high runoff. The levees are made of silt, because there is nothing else to use. There is a limit to the height of earthen embankments; and it is physically, as well as financially, impossible to keep on raising them higher and higher. There is always great danger that they will be breached during times of flood and that the Imperial Valley and other sections will be forced to undergo another inundation.

These objections to a levee system for flood control would

[10] E. O. Leatherwood, "My Objections to the Boulder Dam Project," *The Annals of the American Academy of Political and Social Science*, Vol. CXXXV, No. 224 (January 1928), pp. 133–34.

[11] Weymouth, *op. cit.*, p. 23.

not be so damaging, however, if it could be shown that the flood danger in the lower Colorado River Basin was greatly exaggerated. It should be noted that the Imperial Valley was the only section in immediate danger, and that even there it would be years before the Salton Basin was completely filled. There would be no danger to human life, since the inhabitants would have plenty of time in which to escape and the water would merely flow into the Salton Sea. During low stages it might be possible to return the river to its old channel, and in any event if adequate levee protection were given to the Imperial Valley it would be unnecessary to complete plans for the development of the lower river until further consideration had been given to all of the various aspects involved.

Unfortunately the flood danger was not exaggerated, and any attempt to employ a halfway measure to solve the problem would undoubtedly have ended in failure. The floods from the upper Colorado River Basin are of such long duration that a permanent channel to the Salton Sea might have been created from which it would have been impossible to force the river back to its old course. It is true that there was no great danger to human life, but the destruction of property would probably not have been gradual. In the first place, the value of immovable property would have declined rapidly in the face of certain inundation. In the second place, actual destruction of much of the property would have occurred long before the flood had reached it. Any serious break would undoubtedly have destroyed the main irrigation canals, and the valley would have reverted very quickly to desert. With the canal system cut off, all of the inhabitants would have been forced to leave at once, and it is doubtful whether there would have been sufficient water available to move livestock to safer pastures.[12] Strange as it may seem, the greatest immediate danger to the Imperial Valley was destruction by drought caused by flood, and levees alone were obviously not a dependable method of protecting the valley from such a catastrophe.

[12] House Hearings before the Committee on Irrigation and Reclamation, 70th Congress, 1st Session, *Protection and Development of the Lower Colorado River Basin* (1928), Statement of Charles L. Childers, Attorney for the Imperial Irrigation District, pp. 432–33.

Although levees are not dependable means of flood protection, it will be necessary to maintain some levees on the lower river, even with the Boulder Canyon reservoir in operation, so long as the floods of the Gila remain uncontrolled. It was argued, therefore, that since the levees below Yuma had to be maintained anyway the cost of the project would be a total loss as far as flood protection to the Colorado River delta lands is concerned. This objection is not as serious as might at first appear, however, since the Gila floods are not nearly as dangerous as the floods from the upper Colorado River. It is true that the floods from the Gila were responsible for most of the trouble in 1905 and 1906 and that the river at times carries floods of huge proportions; but these floods are "flashy" in character and of very short duration. Although their sudden appearances and erratic occurrences make them peculiarly menacing to levees, they do not last long enough to carve a permanent channel to the Salton Sea. The quick decline in their volume makes their control relatively easy, and even if the levees were breached the river could soon be returned to its former channel.

As a matter of fact, the destructive floods which in the past have swept down the Gila are no longer to be expected, since developments upstream and on the tributaries have done much to regulate the flow of the stream, and future developments will complete the work. The Parker Dam will solve the flood and silt problem for the Williams River, but it will take more than one dam to achieve similar results for the Gila. In 1920 the Reclamation Service made an extensive study of the Gila River from source to mouth and concluded that the best solution to the problem was the construction of the Coolidge Dam, about 90 miles east of Phoenix.[13] This was only a partial solution, since a number of tributaries enter the Gila below the dam site; but it reduced the flood menace from about 200,000 second-feet to about 100,000 second-feet. The construction of the Coolidge Dam was authorized by an act of Congress on June 7, 1924,[14] and was completed in 1928. The Salt River, the most important

[13] G. E. P. Smith, *The Colorado River and Arizona's Interest in Its Development,* pp. 545–46. This dam was originally known as the San Carlos Dam.

[14] *United States Statutes at Large,* Vol. XLIII, Part 1, p. 475.

tributary of the Gila, was partially regulated as early as 1911 with the completion of the Roosevelt Dam.[15] The building of the Mormon Flat Dam in 1925, the Horse Mesa Dam in 1927, and the Stewart Mountain Dam in 1930 placed the Salt River under complete control. The Verde River, a tributary of the Salt, will be regulated by the Bartlett Dam, and Cave Creek, a tributary of the Verde, was placed under control with the completion of the Cave Creek Dam in 1923. The Aqua Fria, a tributary of the Gila below the Salt, is regulated by the Lake Pleasant Dam; and the Hassayampa, another tributary, will probably be dammed at Walnut Grove by private enterprise. Plans are also under consideration to construct a dam at Charleston on the San Pedro River, the only remaining tributary of the Gila that might cause flood damage.[16] With all of these developments on the Gila and its tributaries it does not appear unreasonable to conclude that the recurrence of disastrous floods on this river is very unlikely. In view of the facts that the Gila floods were always of a very "flashy" nature and that the extremely high peaks will never occur again, the statement that the Boulder Canyon Project is an uneconomic and ineffectual method of flood control on the Colorado River because no provision was made to regulate the Gila appears to be unwarranted.

It was decided, therefore, that to dam the Colorado in one of its canyons was the best method to achieve flood control. But there was still the question whether or not a smaller reservoir than Lake Mead located at any of several different possible sites would be a more economic way of reaching the same results. Floods in the lower Colorado come from three sources: (1) the headquarters of the Green, the upper Colorado, and the San Juan, (2) the intermediate area between the San Juan and the Gila, and (3) the Gila Basin. It has already been shown that the construction of storage reservoirs on the tributaries in the upper basin would not be the most economic method of regulating the river, since these reservoirs would not intercept the

[15] E. C. La Rue, *Water Power and Flood Control of Colorado River below Green River, Utah* (1925), p. 34.

[16] House Report No. 1657, 69th Congress, 2d Session, *Boulder Canyon Reclamation Project*, Part 3 (1927), p. 26.

floods of tributaries below the San Juan, they are too far away from the flood areas to give effective control, and most of the dam sites are better suited to power production than to water storage. Of the sites in the intermediate area, the Black Canyon site was found to be the best suited for the first great multiple-purpose project in the lower Colorado River Basin, although for flood-control purposes alone the Mohave site deserved serious consideration. It was stated that flood control was the only urgent need and that there was no demand for increased irrigation development and increased power production in the Southwest. Therefore the first step should be to provide flood-control works quickly and at a minimum cost without destroying other sites that are valuable for other uses.[17] A storage of not more than 10,000,000 acre-feet would be needed; and the Mohave site is well suited to this purpose, although it should be noted that, like Black Canyon, it is also located above the Gila and Williams rivers and would not control the floods in those streams. A reservoir of this size could be built for much less than the cost of the Boulder Canyon Project, and it was urged that a smaller reservoir for flood control only should be constructed as a substitute plan. It was argued that such a reservoir could be built at a cost not to exceed $15,000,000 and that an expenditure of $165,000,000 for the Boulder Canyon Project was thus beyond all reason. It was stated that the property to be saved was not worth over $100,000,000, and that it was very poor business to spend an amount in excess of the value of the property to be saved, especially when a much less expensive plan to achieve the same result could be followed.[18]

On the surface this argument appeared to be a reasonable statement of the situation; but it neglected certain important facts which placed it in an entirely different light. The argument that the property to be saved in the lower basin is worth less than the cost of the Boulder Canyon Project involved a misstate-

[17] House Hearings before the Committee on Irrigation and Reclamation, 68th Congress, 1st Session, *Protection and Development of Lower Colorado River Basin* (1924), Testimony of Mr. E. C. La Rue, Hydraulic Engineer, United States Geological Survey, p. 969.

[18] House Hearing before the Committee on Rules, 70th Congress, 1st Session, *Boulder Dam,* Part 2 (1928), Statement of Hon. Lewis W. Douglas, pp. 53–54.

ment of fact[19] and overlooked the great possibilities for future developments and increasing values in that section. If a plan costing $15,000,000 would achieve the same results, however, it was obviously foolish to spend $165,000,000. But a reservoir of some 10,000,000 acre-feet capacity could not possibly be as satisfactory a solution to the flood-control problem as the larger reservoir in Black Canyon. Only $25,000,000 of the $165,000,000 has been allocated as a flood-control burden, and this allocation is admittedly an arbitrary one, as is necessarily the case under conditions of joint cost, although it appears to be a reasonable figure in view of the fact that the Boulder Canyon Project provides other services which are able to bear a share of the cost burden. The smaller reservoir would provide no space for silt storage and would soon lose its effectiveness in flood control. With no upstream developments the reservoir capacity would have been reduced from 10,000,000 acre-feet to 8,000,000 in less than fifteen years and the whole reservoir would have been filled with silt in seventy or seventy-five years. There could have been very little storage for irrigation purposes, and no power could have been developed because the reservoir head would have fluctuated too violently as the flood waters were caught and then discharged. Thus the requirement of the federal government that the project should provide means of repayment of its cost would not have been fulfilled, and the flood control function would not have been adequately performed by the smaller reservoir.

It is doubtful if a reservoir of 10,000,000 acre-feet capacity could have been built for $15,000,000, and in any event the least expensive way to secure effective flood control was to reserve adequate space at the top of a multiple-purpose reservoir for that service. In the case of the Boulder Canyon reservoir the construction costs per acre-foot of reservoir capacity would decline up to 34,000,000 acre-feet, and the top 9,500,000 acre-feet of space reserved for flood control could thus be allotted a relatively small share of the project's cost.

Aside from the cost aspects, the large Boulder Canyon reser-

[19] House Hearing before the Committee on Rules, 70th Congress, 1st Session, *Boulder Dam,* Part 2 (1928), Statement of Hon. Philip D. Swing, p. 73.

voir was found to fit into the scheme for the most economic and fullest development of the Colorado River, in spite of the contentions of the opposition to the contrary. The most important features to be considered in the plan for the future development of the river were flood control, water storage for irrigation and domestic uses, and power production within transmission distance of a power market. Surveys have shown that although Lake Mead will submerge some dam sites it will not interfere with any proposed irrigation or power projects and it is so located that it can be used effectively for water storage and power production as well as for flood control.[20] Adequate flood control might have been secured at a lower total cost; but, when all of the services rendered are considered, it is doubtful whether any other scheme of development could have achieved the same results as economically as the Boulder Canyon Project.

The flood-control work of the project was started more than a year before the dam was completed. The regulation of the flow of the Colorado River was begun on February 1, 1935, when the 50 × 50-foot steel bulkhead at the Arizona outer-diversion-tunnel inlet was lowered and the river was forced into the Nevada outer-diversion tunnel under the control of slide gates in the tunnel plug.[21] The threatened destruction to the downstream irrigation projects by flood or drought that previously had been ever present was now practically eliminated, and the silt load that had formerly been deposited in canals and ditches, costing millions of dollars for removal, was greatly reduced. By July 1935 the reservoir contained 3,875,000 acre-feet of water, covered an area of about 37,700 acres, and was about 78 miles long. During that spring a flood of 105,000 second-feet was recorded in the river above Black Canyon which might have caused great damage to the Imperial Valley if it had reached the delta region. The levees along the lower river were cracked as a result of an extended drought, but the fact that Hoover Dam was ready to do its flood-control work saved the considerable cost of levee repair.[22] In the spring of 1936

[20] Weymouth, *op. cit.*, p. 25.

[21] *The Reclamation Era*, Vol. XXV, No. 9 (September 1935), p. 180.

[22] *Annual Report of the Secretary of the Interior for the Fiscal Year Ended June 30, 1935*, pp. 46, 53.

the reservoir filled to one-seventh of its capacity and became the largest artificial body of water in the world. By November 1939 it contained nearly 25,000,000 acre-feet of water, and by July 1940 Boulder storage was 32,360,000 acre-feet, or 1,860,000 more than the estimated capacity of 30,500,000.[23] An exceptionally heavy runoff occurred in September 1939, when the flow from the upper basin was augmented by a late summer rain which reached cloudburst proportions over a wide area in the lower basin. Releases through the outlets of the dam were cut immediately to the flow necessary for power generation, and both the Parker and Imperial dams, although not built primarily for flood control, were used to moderate the flood peak. The river flow, which might have become a raging flood repeating the disasters of previous years, was thus held to 35,000 second-feet—not a particularly dangerous height.[24]

Rumors that the presence of this large body of water would influence weather conditions in the Southwest have proved false, since not enough water evaporates from the reservoir to affect the climate noticeably.[25] With the storage of large quantities of water, however, the reservoir can regulate the flow of the lower river for irrigation as well as for flood control, and a rapid increase in the acreage under cultivation in the lower basin is anticipated. The Boulder Canyon Project has thus not only saved from destruction the existing agricultural empire along the lower Colorado but has also extended its possible boundaries to include lands which had previously been considered beyond all hope of reclamation from the desert wastes.

WATER FOR AGRICULTURAL USE

The flood menace was not the only danger faced by the agricultural areas of the lower Colorado River Basin. Destruction by drought was almost as great a peril. The rains are infrequent and insufficient, and the comparatively few wells that exist are not reliable. The climate is so dry that within a few days plants

[23] *Engineering News-Record,* Vol. CXXV, No. 4 (July 25, 1940), p. 123.
[24] *The Reclamation Era,* Vol. XXIX, No. 11 (November 1939), pp. 317–18.
[25] *Ibid.,* Vol. XXVII, No. 1 (January 1937), p. 8.

and animals alike would die if the waters of the Colorado failed to reach them. For this reason the development of the basin had been limited to the acreage irrigable by the average low flow of the stream, and thus during periods of high flow, millions of acre-feet of water had been allowed to run unused to the Gulf. With the enormous storage space available in the Boulder Canyon reservoir, however, the flood waters may be conserved, the flow of the river may be regulated, the periods of drought may be eliminated, and the areas of land placed under irrigation may be greatly extended. These advantages to agriculture were realized soon after the settlement of the Imperial Valley was begun, and plans were suggested for the construction of storage works to secure the fullest development of the resources of the lower basin. During those early days, unfortunately, the task proved to be too difficult and too expensive for private enterprise and the federal government was not as yet disposed to carry on the work.

The need for water storage to prevent drought was demonstrated as early as 1902, which was a year of exceptionally low flow. Since reclamation in the lower valleys had just begun at that time, the irrigated area was small and no actual shortage occurred. Nevertheless, the fact that the river's volume had reached such a low point indicated that losses from drought would be incurred during the low flow periods in the future if production were carried much further. By 1915, another year of small runoff, the area irrigated had increased substantially. Although the flow was not as low in that year as in 1902, all of the water in the river was diverted into the Imperial Canal to irrigate the valley lands during the low-flow period, and an actual shortage existed for part of the time. A shortage occurred again in 1919; but improved methods in application of the water then helped to alleviate the serious condition.[26] In spite of this deficiency of water the area of land under irrigation was extended in both the upper and the lower basins, although it was known that the limit set by the average low flow of the river had probably been reached and that cultivation had been extended

[26] Senate Document No. 142, 67th Congress, 2d Session, *Problems of the Imperial Valley and Vicinity* (1922), pp. 5–7.

far beyond the acreage that could be adequately irrigated during the periods of very low flow.[27] In 1924 the first great loss due to drought was sustained in the Imperial Valley. For ninety-six days all of the water in the river was diverted into the Imperial Canal, and after half of the water in the canal had been taken to fulfill Mexican requirements there was not enough left for proper irrigation of the American lands. The crop loss was estimated at not less than $5,000,000, and for several days during the drought there was barely enough water for stock and domestic purposes.[28]

The Boulder Canyon Project solved the drought problem through the provision for water storage sufficient not only for present needs but for considerable expansion in the area irrigated, and undoubtedly it eliminated the threat of lawsuits and injunctions by the lower basin to force the upper basin to augment the flow of water down the river during periods of small runoff. Unfortunately the project was not completed soon enough to prevent one more disastrous drought season. In 1934, the year before storage in Lake Mead was begun, the Imperial Valley suffered a drought damage estimated at $10,000,000 when the lower Colorado virtually ceased to flow for a period of several weeks.[29] If storage could have been begun one year sooner this disaster might have been averted, and its occurrence showed definitely the great need for water storage to protect the lower basin from drought as well as from flood. In 1936 the river flow again reached a low point, but because of the water stored in Lake Mead, no losses were incurred.[30] Drought damage was not likely to be as permanent as damage by flood, especially in the Imperial Valley; but the threat of such losses was another powerful economic argument in favor of the Boulder Canyon Project.

In addition to solving the drought problem the storage of

[27] Smith, *op. cit.*, p. 537.

[28] House Report No. 1657, 69th Congress, 2d Session, *op. cit.*, Part 1 (1926), p. 16.

[29] *Annual Report of the Secretary of the Interior for the Fiscal Year Ended June 30, 1935*, p. 46.

[30] *The Reclamation Era*, Vol. XXVII, No. 1 (January 1937), p. 2.

water in the Boulder Canyon reservoir and the regulation of the river's flow will result in the extension of the area of land under cultivation in the lower basin. The seasonal lows in stream run-off will be greatly modified, and the floods which once flowed unused to the Gulf can now be conserved and applied beneficially in the reclamation of desert lands. It was estimated that 15,000,000 acre-feet of storage would be required for the complete irrigation development of the lower basin, although a smaller amount would suffice for the immediate future. The Boulder Canyon reservoir will not provide the full amount of storage indefinitely because of the accumulations of silt, but with future construction for storage at other sites on the stream, the full amount can be maintained. Eventually about 1,500,000 acres of land will be added to the productive districts of the lower basin, and a great, new agricultural area will come into existence in the lower Colorado region.[31]

The lands of the lower basin were classified as 44 per cent public, 40 per cent private, 1 per cent state, 2 per cent railroad, 8 per cent Indian, and 5 per cent entered.[32] The Boulder Canyon Project Act provides that all of the public lands shall be withdrawn from entry until water can be furnished for irrigation, and it is obvious that the federal government is greatly interested in the possibilities of the farms and homes that may be made available to ex-service men and others when the irrigation works are completed. Offhand it would seem reasonable to conclude that this carefully planned development of the Colorado River Basin as provided in the Colorado River Compact and the Boulder Canyon Project Act will react to the economic benefit of the entire nation; but several arguments were developed which weakened the validity of this conclusion: It was stated that there was no urgent demand for an increase in irrigation development[33] and that actually a surplus of farm prod-

[31] E. C. La Rue, *Colorado River and Its Utilization* (1916), pp. 144, 158–59. See also Department of the Interior, *Development of the Lower Colorado River* (1928), pp. 386–88. See also *The Reclamation Era*, Vol. XXVII, No. 2 (February 1937), p. 27.

[32] *New Reclamation Era*, Vol. XXI, No. 5 (May 1930), p. 89.

[33] House Hearings before the Committee on Irrigation and Reclamation, 68th Congress, 1st Session, *Protection and Development of Lower Colorado River Basin* (1924), Testimony of Mr. E. C. La Rue, p. 969.

ucts was already being produced in the United States. Under these circumstances, it was argued, a project which would create still further competition for the overcrowded markets was economically unsound; and apparently the general farm population of the country shared this opinion. At the sixtieth annual session of the National Grange a resolution was adopted to the effect that no more federal appropriations should be made for new reclamation projects until the farm products already being produced could be profitably marketed. This resolution was also adopted by the Washington and Idaho state granges and indicated that opposition to, rather than support for, the Boulder Canyon Project could be expected from agricultural organizations.[34]

In 1927 a report was made by the Colorado River Fact-Finding Committee of Utah which concluded definitely that there was no need for additions to the productive acreage in the Imperial Valley in California, or in the United States.[35] The market, economic, financial, and even psychological conditions in the Imperial Valley were found to offer no justification for the support or encouragement of a plan to extend the area under irrigation, especially since 190,000 acres of land under irrigation were lying unused in the valley at that time. It was estimated further on the basis of a study made by the University of California that it would take from fifteen to twenty years to develop all of the irrigated land in California for which construction had already been provided[36] and that expansion in that state was entirely unnecessary. For the country as a whole the situation was summarized by Mr. W. M. Jardine, Secretary of Agriculture, in his annual report for 1927 in which he stated that there was no scarcity of agricultural land in the United States. Although the federal government had disposed of practically all the lands of agricultural significance formerly in the

[34] House Report No. 918, 70th Congress, 1st Session, *Boulder Canyon Power Project*, Part 3 (1928), Minority Views by Mr. Leatherwood, p. 25.

[35] Colorado River Fact-Finding Committee of Utah, "*All-American*" *Canal* (1927), pp. 6–7.

[36] David Weeks and Charles H. West, *The Problem of Securing Closer Relationship between Agricultural Development and Irrigation Construction* (1927), p. 3.

public domain, there was still an area of some 600,000,000 acres of potential crop land in private ownership. Thus it would be unnecessary for many years to stimulate agricultural expansion in the United States, because if farming should become sufficiently profitable plenty of land would appear automatically.[37] On the basis of studies and statements such as these, the conclusion was reached that the provision for the enormous water-storage space in the Boulder Canyon reservoir would serve only to accentuate the depressed condition in the agricultural industry and that a much smaller project should be designed to provide flood control and water storage for present needs only.

It is undoubtedly true that a large acreage of land which might have been made productive lay idle in 1927. But this argument alone should not condemn plans for the future expansion of the country's agricultural area. With the normal increase in the country's population more land will be brought into production, and in spite of the depressed condition of agriculture in 1927 and its still poorer situation in 1933 a general need for more agricultural lands may not be as long delayed as the opposition indicated. Apparently it was assumed that the additional lands would be placed under cultivation almost as soon as the Boulder Canyon Project Act was passed; but this, of course, would be impossible. Five or six years would elapse before construction could be completed to the point where water storage would begin, and even then several more years would pass before the lands could be brought into production. Also the new lands would be an important addition to the lower Colorado River Basin but would represent only a very small fraction of the total agricultural area in the United States. This fraction would be so small that the new lands could have only an infinitesimal effect upon the agricultural situation, even when the entire irrigable area was finally placed under cultivation,[38] and it hardly seemed worth while to object to such an addition. It should be noted, however, that the capacity of the Boulder Canyon reservoir reserved for water storage was obviously

[37] *Report of the Secretary of Agriculture, 1927*, pp. 26–27.

[38] Paul T. Homan, "Economic Aspects of the Boulder Dam Project," *Quarterly Journal of Economics*, Vol. XLV, No. 2 (February 1931), p. 208.

meant to promote the extension of the cultivated area in the lower basin and that if an expansion of agricultural area in the near future was in itself economically unsound a valid objection to the project on this ground existed, even though the increase in area would be delayed a few years and the acreage would be comparatively small. Actually the basic question to be decided was not whether the extension would be delayed or how great it would be but whether or not the cultivation of the specific lands in question would react in any way, great or small, to the economic benefit or detriment of the country.

In the writer's opinion the lower Colorado River Basin as a definite agricultural development should not be compared with the agricultural industry of the country as a whole, since the chief crops there do not compete with the chief crops in most other sections of the nation. In the lower basin the summers are hot and the winters are mild. The principal crops are lettuce, peas, spinach, other vegetables, cantaloupes, grapes, citrus fruits, and strawberries. The warm climate makes it possible to produce these products at an early season, and they appear on the Eastern markets at a time when they do not compete seriously with similar products from other parts of the country.[39] The lower-basin source of supply makes it less difficult to obtain these types of food during certain seasons and thus to some degree adds to the variety of consumption and the satisfaction of the consumers. It is true that cotton is produced in the Imperial Valley; but most of it is grown on the Mexican side of the border where the land is held in larger tracts and a different labor situation exists. On the American side a greater return can be secured from diversified vegetable and fruit crops, and the products from the new American lands will undoubtedly be of that character.[40] The Western markets for the products of the lower basin are expanding very rapidly, especially in southern California; and the anticipated influx of capital and population in that region will probably expand these markets more rapidly

[39] House Report No. 1657, 69th Congress, 2d Session, *op. cit.* (1926), pp. 14–15.

[40] E. F. Scattergood, "Engineering and Economic Features of the Boulder Dam," *The Annals of the American Academy of Political and Social Science*, Vol. CXXXV, No. 224 (January 1928), p. 122.

than cultivation can be extended on the lower Colorado River lands to meet the new demand. As a result the little competition that exists in the Eastern markets is likely to decline still further rather than increase with the agricultural development of the lower basin. Thus it is not the extension of the total agricultural area alone which should be the determining factor in an economic analysis of the irrigation services rendered by the Boulder Canyon Project. The climate, the character of the crops, and the location in relation to markets are considerations of even greater importance; and these considerations point to the conclusion that the development of the lower Colorado River Basin will add to the economic well-being of the people and will not detract from the prosperity of any other branch of the agricultural industry.

Even with the advantages of the lower-basin lands in product, climate, and location, the question still arises whether or not the values created in the extension of the area of irrigable lands are worth the expense incurred in creating them. It is impossible to forecast accurately what the values will be; and it is equally impossible to find anything but an arbitrary figure for the cost of a service produced under conditions of joint supply. The cost of the All-American Canal will be borne by the lands served by the canal; but none of the costs of the Black Canyon structures are to be repaid by the water stored for agricultural use.[41] On such a basis, there is no doubt that the land values created will be greater than the costs attributed to them, although the entire picture might be changed if the cost burden were placed on irrigation rather than on power production and water for domestic use. However, the lands of the lower Colorado River are rich lands, and even when the agricultural industry as a whole was in the depth of depression communities in this section forged ahead. Such lands are of great value to the United States, and the creation of more of them must necessarily help to improve the economic condition of the country. Until recently the Colorado River was a great national resource which was being wasted. Now it is being ex-

[41] See Table XI, p. 184, above.

ploited according to a well-devised plan and will be used to increase the wealth and further the prosperity of the Colorado River Basin and, to a lesser extent, of the entire country.

The lower Colorado River lands are extremely fertile, but it has been argued that they are expensive lands to cultivate.[42] This statement was based on the observations that the hot climate would cause great evaporation and that the unused share of the water taken to the Imperial Valley could not be returned to the river and would flow as waste to the Salton Sea. Utah claimed to have even more fertile lands which would be cheaper to cultivate if the waters of the Colorado could be brought to them; but probably these lands are not as well located in relation to markets or do not enjoy as good a climate as the lands of the lower basin. The most productive lands tend to be developed first, and what the lower basin lacked in fertility and in thrift of water application was more than overcome by the location of the lands and the early and long growing season. With the opening of new productive areas in this region some movement of population from the poorer lands elsewhere in the country to these better lands may be expected, although in relation to the total farming population this movement will probably be too slight to be readily noticeable. Such a movement is beneficial from an economic point of view, since it tends to increase the productivity of the labor and capital applied to the better land. It cannot be considered detrimental, for it is as sound economically as the replacement of an old, obsolete, and unproductive piece of equipment with a new, effective, and highly productive machine, or the abandonment of Atlantic seaboard farms for the more fertile lands in Iowa. To continue the use of an inefficient economic agent instead of an available, highly productive substitute would be to disregard economic principles rather than to follow them. The most damaging economic argument against the water-storage phase of the Boulder Canyon Project is that the water will undoubtedly be reserved for use on California and Arizona lands

[42] House Hearings before the Committee on Irrigation and Reclamation, 70th Congress, 1st Session, *Protection and Development of the Lower Colorado River Basin*, Part 2 (1928), Statement of Hon. George H. Dern, Governor of Utah, p. 197.

some of which are not as productive as the delta lands on the Mexican side of the border; but, unfortunately, as between nations, economic principles are often disregarded in order to serve the dictates of political expediencies.

In addition to water storage in Lake Mead for agricultural use, the Boulder Canyon Project provided for the construction of the All-American Canal, which was designed to head at Imperial Dam and to convey water entirely within the United States to the Imperial and Coachella valleys. The capacity of the canal was great enough to supply lands for which no water had been available in the old Imperial Canal; and, since it headed at a higher elevation, it would reach some sections by gravity which previously could have been reached only by pump lifts. The All-American Canal will extend greatly the irrigable area in the Salton Basin and will provide a more reliable supply to those American lands already under cultivation. It should be noted, however, that the original motivating force behind the All-American Canal idea was political rather than economic; and the political aspects are still the dominating considerations. It was the desire of the Imperial Valley to free itself of Mexican interference which prompted the construction of a canal entirely within the United States. Even the name "All-American" indicates the strength of this motive. The economic aspects of this phase of the Boulder Canyon Project were thus of secondary importance, and although they are worthy of careful consideration the justification of the canal is not on as firm a foundation economically as the justification of the other features of the project.

Before construction of the All-American Canal was begun it was pointed out that the Imperial Canal was still in excellent condition and that it had the advantage of following the natural route from the Colorado River to the Salton Basin. It would have been much less expensive to improve and enlarge this waterway than to build the new canal; and although a smaller area would then be reached by gravity, most of the additional irrigable lands in the Imperial Valley could be reached by pump lifts. The All-American Canal would add nothing to the supply of water available in the river and might actually cause a loss of water through seepage and evaporation.[43] Its estimated cost of $38,500,000 would be a heavy burden on the lands benefited, and some of those whose lands were already being served by the Imperial Canal objected to the increased taxation that the new canal would involve. The indebtedness of the Imperial Irrigation District already amounted to $16,000,000, or about $30 per acre, and the All-American Canal would add about $40 to that, making a total of $70 per acre. This amount would have been even greater if the entire cost of the new canal had been allocated to the Imperial lands already under production; but a large share of the cost would be borne by the Coachella Valley and by the additional lands placed under production in both districts. However, the spreading of the cost burden to the new lands would be largely offset by the loss of the contributions from the Mexican lands. The Imperial district would abandon the old canal, a $7,000,000 project, and its treasury would no longer receive the $600,000 or so per annum from the Mexican lands in payment for the use of water.[44] Yet in spite of the abandonment of the Imperial Canal and the loss of revenue the Imperial Valley would not be relieved entirely of Mexican interference, since some protective levees would still have to be maintained on Mexican territory,[45] and the political objective of complete freedom from foreign influence would not be achieved.

The arguments were also advanced that the canal project was not sound from the engineering point of view and that the cost would probably be double or triple the estimates. These last statements were merely conjectures, however, and have since been disproved. Nevertheless the higher-cost argument was used in an attempt to show that the lands benefited could not bear the burden of the cost of the canal, especially in view of

[43] William Kelly, "The Colorado River Problem," *Transactions of the American Society of Engineers*, Vol. LXXXVIII (1925), Paper No. 1558, discussion by J. C. Allison, construction engineer, Calexico, California, p. 353.

[44] Colorado River Fact-Finding Committee of Utah, *"All-American" Canal* (1927), pp. 7–8.

[45] Senate Hearings before the Committee on Irrigation and Reclamation, 69th Congress, 1st Session, *Colorado River Basin*, Part 2 (1925), Statement of Charles E. Scott, Representing the Colorado River Control Club, pp. 195–97.

the fact that agricultural prices had fallen since the World War period, and that eventually the government would have to foreclose on the valley and become a permanent landlord.[46] Since this conclusion was based upon inflated cost figures it was obviously farfetched, and although the evidence to answer the question whether or not the Salton Basin was able to bear the increased burden was inconclusive the investigations on that subject indicated definitely that the answer would be in the affirmative.[47]

It was true that the old Imperial Canal could be enlarged and improved much more cheaply than the new canal could be built, and it was also true that water (if available) could be pumped from the old canal to the additional lands which would be served by the All-American Canal. It should be pointed out, however, that under a pumping system, as a practical proposition, this condition would never be achieved. Much of the additional land in the Imperial Valley could not bear the heavy pumping charges involved, and under such a system it would undoubtedly have been too expensive to bring any Colorado River water at all to the Coachella Valley. If those fertile lands were to be developed, the construction of an all-American canal which would enter the Salton Basin at a higher level than the Imperial Canal was the only practical solution to the problem. Obviously, the additional lands in both the Imperial and the Coachella valleys could not bear the cost of an all-American canal alone if they could not bear the smaller total burden of pumping charges; but, fortunately for these lands, the areas already under cultivation in the two valleys were willing to assume a large share of the cost of the new canal in order to escape Mexican interference and to secure a more reliable source of supply.[48] It was estimated that with the All-American

[46] House Hearings before the Committee on Irrigation and Reclamation, 70th Congress, 1st Session, *Protection and Development of the Lower Colorado River Basin* (1928), Appendix, Letter to Representative Carl Hayden of Arizona from Epes Randolph, pp. 568–69.

[47] House Document No. 359, 71st Congress, 2d Session, *Report of the American Section of the International Water Commission, United States and Mexico* (1930), pp. 113–16.

[48] House Report No. 1657, 69th Congress, 2d Session, *op. cit.*, Part 1 (1926), p. 14.

Canal well over 200,000 acres of land would be placed under irrigation which otherwise might never be cultivated.

After the construction of the new canal and the Imperial Dam it would no longer be necessary to build temporary weirs in the river during low-flow seasons in order to raise the water sufficiently to make diversions. This expense would thus be avoided, although in itself it was not a sufficiently large item to carry much weight toward the economic justification of the canal. It was claimed that the health of the people in the American section of the Imperial Valley was constantly imperiled, since the Mexicans near the old canal threw everything they did not want into it, including dogs, horses, cattle, human corpses, and all sorts of refuse. Because this water was used for potable purposes as well as for irrigation, an epidemic (of typhoid fever, for instance) which might start below the border would spread quickly to the American farmers. Although this danger was probably very real, it should be noted that undoubtedly there are other much less expensive ways to safeguard the health of the American communities in the Imperial Valley than to construct a canal entirely north of the border. It was also claimed that there was a serious noxious-weed situation in Mexico and that within a few years the plants would follow the banks of the main canal into the United States.[49] The undesirable possibilities of such a condition were another argument in favor of an all-American canal; but again it should be noted that there are much cheaper methods of weed control than the complete separation of the Mexican and American irrigation systems in the Imperial Valley. If the canal is to be justified economically, it must be shown that it will bring so much more land under irrigation by gravity or with smaller pump lifts, that the saving involved is greater than the cost of the canal plus interest, and that this method is more economical than any other possible method of achieving the same results. It was pointed out that it would be cheaper to build a small all-American canal to serve the high Imperial lands and the

[49] Senate Hearings before the Committee on Irrigation and Reclamation, 69th Congress, 1st Session, *Colorado River Basin,* Part 2 (1925), pp. 257, 296, 297.

Coachella Valley and to maintain the Imperial Canal for the area already under cultivation. Although the so-called "fixed" costs of construction would have been about the same for a small canal as for a large one, the difference in the variable costs of such projects would have amounted to millions of dollars. Since an irrigation system of this type does not seem to be an unreasonable solution to the problem if political considerations are disregarded, and since the arguments in favor of the new canal have failed to account for the savings involved in this alternative method, the evidence presented to establish the economic justification of the All-American Canal now under construction appears to be inconclusive.

The use of Colorado River water on the Mexican lands is another point for consideration closely related to the All-American Canal phase of the Boulder Canyon Project. It was stated that the regulation of the flow of the river would increase the supply of water in Mexico during the dry seasons and would permit the extension of irrigation of Mexican lands. On the other hand, it was argued that the All-American Canal would provide facilities for the rapid expansion of cultivation on the American side of the border and thus for the increased use of water in the United States. At intervals the flow of water around Hoover Dam could be so regulated that the supply reaching Mexico would be large enough to support only the present cultivated area. Actually the United States, if she so desired, could destroy all Mexican cultivation on the Colorado River delta merely by cutting off entirely the water supply for a sufficient length of time. However, it was very unlikely that the flow of the river would ever be deliberately cut off or even held to the minimum. Not only would such action give cause for international controversy but it would prevent the maximum production of Boulder power upon the sale of which the chief financial burden of the project rested. It was suggested that the problem of keeping the power plant in continuous operation and at the same time preventing the equated flow from reaching Mexico might be solved by diverting the entire river into the All-American Canal, and dumping the surplus into the Salton Sea until the time that the entire flow of the river could be used to irrigate American lands. But aside from the fact that such a procedure would involve a great waste of valuable water, the suggestion ignored the physical characteristics of the Salton Basin. With such a tremendous quantity of water flowing into the basin, the Salton Sea would rise rapidly and submerge farm lands and other valuable property. The Imperial Irrigation District had even entered a contract with the Southern Pacific Railroad Company by which the district promised not to permit the level of the water to rise more than three feet, since beyond that point the railroad's property would be endangered.[50] It was well known that this storage space would be occupied very quickly if the entire river were diverted into the basin. There seems to be little doubt, therefore, that the equated flow of the river will be allowed to reach Mexico; and it is obvious that the Mexican farmers can easily build a new heading for the Imperial Canal on Mexican territory to use the old irrigation system as before and even to extend their acreage under cultivation. Nevertheless the advantages gained by Mexico through the regulation of the river's volume should not cause great alarm to the American agricultural interests. As previously explained, according to the principles of international law the water in the stream belongs to the United States until it enters Mexican territory, and in the absence of an agreement with Mexico, the United States has the right at any time to take as much of the water as she desires. An agreement with Mexico now would avoid possible hard feelings in case of an argument in the future; but the Mexican farmers should realize that some day the United States might make use of the equated flow to irrigate her own lands and that Mexican cultivation beyond the present acreage will then probably lack water sufficient to maintain the extended area in a productive state.

The state of Arizona was more alarmed than any other section of the lower basin by the possible extension of the cultivated area in Mexico. With the All-American Canal, California could increase her acreage as rapidly as Mexico; and Nevada did not

[50] House Hearings before the Committee on Irrigation and Reclamation, 70th Congress, 1st Session, *Protection and Development of the Lower Colorado River Basin,* Part 2 (1928), Statement of Hon. George H. Dern, Governor of Utah, pp. 200–201.

have enough agricultural land subject to reclamation under the Boulder Canyon Project to be seriously interested in the situation. Arizona at one time asked for the construction of a "highline" canal to make diversions from a point far enough up the Colorado River to bring water to the high Arizona lands by gravity. Arizona maintained that such a canal would permit the rapid extension of her cultivated area; but the investigations showed that the cost of the project was too great to be within the limits of economic practicability.[51] Naturally Arizona is greatly concerned with the development of the lands within her borders; but from an economic point of view it would be an error to stimulate the development of Arizona lands while more productive lands in Mexico lie unreclaimed. It was feared by some that low-cost Mexican products would steal the markets belonging to the American farmers and that such a condition would be disastrous to the American section of the Imperial Valley. Such fears were probably greatly exaggerated; but even if such a condition should some day obtain it may be economically beneficial to society in the long run through a lowering in food costs. In general it may be said that the storage of water in the Boulder Canyon reservoir for agricultural use is economically justifiable, since it permits the cultivation of an area more productive than most other agricultural lands in the country. That some of the best lands in the lower basin may not be developed merely because they happen to be in Mexico is an unfortunate circumstance; but this condition is a result of a political situation and can probably never be reconciled to the fundamental economic principles involved.

WATER FOR DOMESTIC USE

The Colorado River is the lower basin's chief source of water supply for potable purposes as well as for irrigation; and as the population of this section of the Southwest increased, more and more water was taken from the river for domestic uses. The amount of water taken for potable purposes, however,

[51] Smith, *op. cit.*, pp. 543–45.

appeared insignificant when compared with the amount diverted for irrigation. It was known that without drinking water from the Colorado a large part of the population of the lower basin would be forced to leave that district immediately; and the need for sufficient water to maintain the personal health of the inhabitants was thus universally recognized. The chief objections to diversions for potable purposes centered on the Colorado River Aqueduct project under which over 1,000,000 acre-feet of water per annum would be diverted from the basin to the coastal plain to be used by the cities comprising the Metropolitan Water District. Objections would probably have been made also to the proposed diversions for the cities of Denver, Colorado, and San Diego, California, if definite steps had been taken to build the structures necessary for such diversions; but under the circumstances the City of Los Angeles, as the most important member of the Metropolitan Water District, became the obvious target for the opposition arguments.

For many years the City of Los Angeles had been searching for a reliable water supply to meet the needs of its rapidly increasing population. The construction of the Los Angeles Aqueduct to the Owens River in the Sierra Nevada Mountains was a great feat of engineering and had received world-wide publicity. Because of this previous search it was generally accepted that the Los Angeles area was in great need of water, and the city's proposal to build another aqueduct to the Colorado River was not an unexpected development. Since regulation of the river's flow and the availability of a large supply of cheap power were essential to the successful operation of such an aqueduct, the provision of a domestic water supply for certain districts in southern California was a strong argument for securing the enactment of the Boulder Canyon legislation. With the development of this argument the need of the Los Angeles area for a larger water supply was placed under closer scrutiny, and some doubts were expressed concerning the urgent necessity for an augmented supply. The capacity of the Los Angeles Aqueduct is 400 second-feet, which will provide 100 gallons per day per capita (not counting losses) to about 2,585,000 people. Diversions from the Colorado of 1,500 second-feet would provide

100 gallons per day to an additional 9,694,000 people.[52] The total population that could be supported, therefore (not counting local sources of supply), would be about 12,279,000 people. It was stated that a scheme to provide water for that number was utterly fantastic, especially in view of the fact that a population of only 2,200,000 occupied the area to be served by the new aqueduct.[53] If the proposed supply was unreasonably large, it was evident that it would be far sounder economically to use the water on the fertile lands in the Colorado River Basin to supply the growing needs in southern California than to supply that district with a domestic water supply for a population which does not now and may never exist there.

The weakness in this argument was that a rapid increase in the population of southern California to 10,000,000 or more was not as fantastic as might at first appear, and it could be proved definitely that when the population of the Los Angeles area grew to about 2,500,000 it would have reached the limit of the water supply available from the Los Angeles River and the Los Angeles Aqueduct. Either a new supply would have to be found, or the growth in population would have to cease. In 1933 it was estimated that the city's population was increasing at the rate of about 100,000 per annum and that the limit of the water supply would be approached in about ten years.[54] Between 1920 and 1930 the city's population increased 114.7 per cent, and the population of Los Angeles County increased 135.8 per cent.[55] Statistics showed, therefore, that there was a tremendous growth in population in the Los Angeles area; and there was no reason to assume that the rapid growth would not continue as long as the requisites for healthful living were avail-

[52] One second-foot = 646,272 gallons per day; 400 × 646,272 = 258,508,000; 1,500 × 646,272 = 969,408,000. See John Clayton Hoyt and Nathan Clifford Grover, *River Discharge* (1916), p. 205.

[53] In 1930 the population of Los Angeles County, which included most of the municipalities composing the Metropolitan Water District was 2,208,492. See *Fifteenth Census of the United States, 1930, Population,* I, 131.

[54] House Hearings before the Committee on Irrigation and Reclamation, 68th Congress, 1st Session, *Protection and Development of Lower Colorado River Basin* (1923), Letter of the Board of Public Service Commission of the City of Los Angeles, p. 331.

[55] *Fifteenth Census of the United States, 1930, Population,* I, 128, 131.

able. During 1938 the population of the metropolitan area probably passed the 2,500,000 mark; and it was forecast that according to present trends the population of this district would reach 7,500,000 by 1980.[56]

That part of California lying south of the Tehachapi Mountains embraces about 20 per cent of the area of the state favorable for human habitation but has only about one per cent of the state's waters exclusive of the Colorado River. Yet this district was supporting in 1930 a population of some 3,100,000, or 57 per cent of the total for the state. This development could not have occurred without the existence of enormous underground reservoirs in the porous floors of the valleys which had accumulated tremendous quantities of water throughout the ages. These reservoirs had been drawn upon to supply the needs of the present generation, and in spite of some replenishment during years of above normal precipitation the reservoirs were being depleted rapidly.[57] The water levels had dropped at an alarming rate since 1916, the end of a period of wet years; and if an additional supply could not be obtained, there would necessarily be a serious shrinkage in the use of land in this area. A study made in 1924 showed that a supply of 140 second-feet could be obtained from the Los Angeles River Basin and a supply of 400 second-feet from other sources in Los Angeles County, which, with the 400 second-feet from the Los Angeles Aqueduct, would make a total of 940 second-feet. It was estimated, however, that to serve the metropolitan portion of the county until 1933 without overdraft would require 1,315 second-feet and that by 1950 a total of 1,870 second-feet would be necessary. Without an additional supply of water, therefore, an overdraft of 930 second-feet would have been sustained before the second half of the century had begun. Actually, during the summer months of 1930 the City of Los Angeles consumed daily twice the volume of water delivered through the aqueduct;[58] and it is not surprising in view of the estimated figures,

[56] *The International Engineer*, Vol. LXXIII, No. 5 (May 1938), p. 149.

[57] Franklin Thomas, "Metropolitan Water Distribution in the Los Angeles Area," *The Annals of the American Academy of Political and Social Science*, Vol. CXLVIII, No. 237 (March 1930), Part 2, pp. 6–9.

[58] *Engineering News-Record*, Vol. CV, No. 11 (September 11, 1930), p. 429.

therefore, that the Metropolitan Water District should have actively promoted a plan to take 1,500 second-feet from the Colorado River in order to provide for the pressing needs of its constituent municipalities.

Nearly all of the studies of the hydrology of southern California had shown that there is a great scarcity of water in that section of the state, and the California interests had little difficulty in refuting successfully the argument that there was no need for an additional water supply in southern California. However, even if the need had existed, it still had not been proved that the Colorado was the only source of supply or the best source under the circumstances. Statistics showed that Owens Valley in conjunction with Mono Basin could yield about 834,000 acre-feet of water annually,[59] which would amount to 585,000,000 gallons daily, allowing 80 per cent conservation, or enough for a population of 5,850,000 people at 100 gallons per capita. Since this figure was over four and one-half times as great as the population of the City of Los Angeles in 1930,[60] it was contended that the city's present source of supply could be developed to fulfill all of its needs, and that it was entirely unnecessary to go to the Colorado River for water. A second aqueduct as large as the Los Angeles Aqueduct, it was said, could be built to the Sierra Nevada Mountains to bring the additional supply to Los Angeles, and if proper reservoirs were built the water resources of southern California could be greatly augmented through the conservation of local floods. The growth of metropolitan Los Angeles would diminish the horticultural belt now being served through irrigation, and, since for any given area the amount of water needed for the cultivation of citrus fruits is practically the same as the amount required for urban use, the water now consumed for irrigation would be sufficient to care for all domestic uses as the city encroached upon the present citrus belt. In addition, as the city grew, the per capita consumption of water might be expected to decrease;

[59] State of California Department of Public Works, *Flow in California Streams*, Table 3, facing p. 72.

[60] The population of the City of Los Angeles was 1,238,048 in 1930. See *Fifteenth Census of the United States, 1930, Population*, I, 141.

and it was estimated that for a city of 10,000,000 people the daily per capita consumption would be about 75 gallons rather than 100. Thus it was argued that with the complete development of the Owens Valley, Mono Basin, and local sources of supply, enough water was available for a population of 8,000,000 people. It would be economically unsound, therefore, for Los Angeles to take water from the Colorado River and, as a consequence, to deprive some Arizona lands of any possibility of cultivation. In addition it was well known that Colorado River water is hard and dirty and that it cannot be used satisfactorily or safely without a considerable expenditure for chemical treatment. The water from the Sierra, on the other hand, is wholesome, soft, clear, and unadulterated; it would require no treatment for purification except for possible occasional chlorination; and it would reach Los Angeles by gravity.

Under such conditions, it would appear reasonable to conclude that a water supply from the Sierra would be a much cheaper, as well as a much better, supply for the City of Los Angeles than water from the Colorado River. Moreover, the Colorado River Aqueduct would be very long; it would involve heavy tunneling and other construction costs; and because of the high pump lifts it would be expensive to operate.[61]

Actually, however, it was doubtful whether the alternative sources of supply were available or, in some cases, whether they really existed. The local sources were being developed very carefully, but even with systematic conservation the local supply could not be increased greatly. More water might be taken from the Owens Valley and the Mono Basin; but the several investigations which had been made of this territory had failed to reveal who owned the water. Four or five different parties claimed the supply, but no determination of the true ownership had been made.[62] In any event, it was known that the additional amount was limited to 300 second-feet, or enough to fill an aqueduct three-fourths the size of the one already built

[61] Colorado River Fact-Finding Committee of Utah, *Water for Los Angeles* (1927), pp. 3–5.

[62] Senate Hearings before the Committee on Irrigation and Reclamation, 69th Congress, 1st Session, *Colorado River Basin*, Part 1 (1925), Statement of William Mulholland, p. 110.

to the Owens River. This additional water would relieve the situation for a few years, but would not be a permanent solution of the problem.[63] It was not considered worth while to build an aqueduct for this amount of water, since before long another source of supply would have to be found and probably the Colorado River Aqueduct would then be built to fill the need. It would be more economical to build the Colorado River Aqueduct at once and to save the cost of construction and maintenance of a second aqueduct to the Sierra. In addition, if all of the remaining water were taken from the Owens River and the Mono Basin, the lands under irrigation in that territory would be robbed of their water and forced out of cultivation. Such an abandonment of improved lands would represent an economic loss which should be included in the total cost to society of the second aqueduct to the Sierra, if such an aqueduct should be built.

The only other source of supply known besides the Owens River, Mono Basin, and the Colorado River was the Kings River in the San Joaquin Valley; but there again diversions to Los Angeles would cause a reduction in the area of land under irrigation from that river. The diversions from the Owens River into the first aqueduct had resulted in a long and bitter dispute between the city and the Owens Valley farmers, and undoubtedly a similar controversy would take place if an attempt should be made to take water from the Kings River.[64] In spite of all of the legal difficulties which had been encountered by the Boulder Canyon Project, the Metropolitan Water District would probably face fewer injunctions and suits for damages if it took water from the Colorado River than if it attempted to tap the Mono Basin or the Kings River for an increased supply. The comparative proximity of the Sierra and the purity of that water did not overcome the advantages to be gained by securing a permanent right to a reliable supply of

[63] Senate Hearings before the Committee on Irrigation and Reclamation, 70th Congress, 1st Session, *Colorado River Basin* (1928), Statement of H. A. Van Norman, pp. 286, 293.

[64] Senate Hearings before the Committee on Irrigation and Reclamation, 68th Congress, 2d Session, *Colorado River Basin*, Part 1 (1924), Statement of M. S. H. Finley, pp. 146–47.

water from the Colorado River. Although the point is still in dispute, most authorities have agreed that the Colorado is the only dependable source available for a quantity of water large enough to meet the growing needs of the Los Angeles metropolitan area, and on that basis there is little doubt that the construction of the Colorado River Aqueduct would be the most economic method of securing the needed supply.

The citizens of the district are willing to pay for the aqueduct project, and it has been established definitely that they have the wealth to meet this obligation.[65] As a result the construction of the aqueduct practically insures the financial integrity of the Boulder Canyon Project, since the water district must purchase a large block of Boulder power to pump the water to the coastal plain. The sale of power is not the only necessary relationship between the two projects, however, since without the storage and river regulation services rendered by Lake Mead a continuous diversion of water for use in the Metropolitan Water District cities would be impossible. The building of Hoover Dam set the stage for the Colorado River Aqueduct, and unless it can be proved that a better alternative supply exists there is no basis upon which to question the present economic soundness of the plan either from the point of view of the success of the Boulder Canyon Project or from the point of view of the most advantageous development of the Los Angeles metropolitan area.

It will be a number of years before the waters of the lower Colorado are fully utilized, and at the present time it is impossible to state definitely whether or not the rather arbitrary allocations of quantities of water to domestic uses and to agricultural uses are the best divisions that could be made to achieve the maximum benefits to society. Undoubtedly the diversions from the Colorado River Basin to the coastal plain will be very beneficial to the cities of the Metropolitan Water District, and the fact that these cities have joined together to secure the water indicates that the supply will be used to promote the best in-

[65] House Hearings before the Committee on Irrigation and Reclamation, 70th Congress, 1st Session, *Protection and Development of the Lower Colorado River Basin*, Part 1 (1928), Statement of Mr. Swing, p. 14.

terests of the districts rather than wasted or appropriated for unimportant purposes merely to cater to local jealousies or conflicting rights.[66] In general it may be said that an organization such as the Metropolitan Water District, which fosters co-operation among its members rather than rivalry, has vast potentialities for securing the maximum economic use of a scarce resource. However, the question still arises whether or not the diversion of a large quantity of water to the coastal plain is more beneficial to society than the use of that water in the reclamation of desert lands, probably in Arizona. If the water is diverted to the Metropolitan Water District, many acres of desert land will never be reclaimed; but if it is used for irrigation in the lower Colorado River Basin, the growth of the population in southern California must be restricted. The decision to be made, then, is whether the social welfare will be best promoted by the future development of a large rural population in Arizona or by a large urban population in southern California.[67] To the writer's knowledge there are no definite grounds upon which to base such a choice for future policy, and in the case at hand the arguments on both sides seem to run in terms either of prejudice or of self-interest. It should be noted, however, that the domestic use of water is usually considered its most important use and that from the point of view of society land should be denied irrigation if no other supply is available for domestic purposes.[68] It seems reasonable to conclude, therefore, that the reservation of a water supply from the Colorado River for the Los Angeles metropolitan area is in harmony with the best social interests involved. In addition the benefits would be gained for society more quickly, since an increase in population and a development of resources will probably proceed at a much more rapid rate in southern California than in Arizona. Under these circumstances there should be no objections to the diversions from the Colorado to the coastal plain, and the

[66] Franklin Thomas, *op. cit.*, p. 11.

[67] Paul T. Homan, "Economic Aspects of the Boulder Dam Project," *The Quarterly Journal of Economics*, Vol. XLV, No. 2 (February 1931), p. 206.

[68] House Report No. 1657, 69th Congress, 2d Session, *Boulder Canyon Reclamation Project*, Part 3 (1927), p. 26.

construction of the aqueduct should result in the creation of services highly beneficial from both the economic and the social points of view.

It is not an extraordinary thing for a city to provide itself with an adequate water supply, and there is no reason to criticize the Metropolitan Water District for taking steps to safeguard the future of its member cities. The only unusual feature of the plan is that the federal government insisted upon being a party to it; but that is a significant political fact and not one which has any direct bearing upon economic expediency. Present conditions indicate that there is a real need for a larger water supply in southern California and that the Colorado River is the only available source for that supply. Even though the water diverted to the coastal plain might be used to great advantage in the lower Colorado River Basin, its domestic use is apparently a more important use and such diversions are, therefore, a more effective means of utilizing this share of the water. Thus in spite of the fact that the water will be transported out of the natural drainage area of the stream, the result of the construction of the Colorado River Aqueduct is likely to promote rather than to restrict the most economic development of the water resources of the lower basin.

GENERATION OF ELECTRIC POWER

The controversy over the possible economic results of those phases of the Boulder Canyon Project and related developments concerned with the storage and diversion of water for agricultural and domestic uses was chiefly a dispute between the states of Arizona and California. It had been estimated that the combined future demands of the two states for Colorado River water would be considerably greater than the amount available even after storage facilities had been developed,[69] and naturally each of the states was anxious to prove that the greatest benefits would be realized if the waters were used within her particular borders. The controversy over the power aspects of the project,

[69] California Colorado River Commission, *Colorado River and the Boulder Canyon Project* (1931), pp. 201–3.

however, was not a dispute that was confined to the relatively local ambitions of two states. It was a clamorous argument that took on the aspects of a nation-wide debate, chiefly because it involved the whole question of whether or not the federal government should enter into large-scale power production activities, as previously discussed, and because it included still other problems of vital importance to the power industry in general. There is little doubt that the Boulder Canyon Project has set a precedent for direct federal action in the development of hydroelectric power sites, and for this reason the economic success or failure of the Boulder power plant will have great influence in determining the extent of federal activity in future projects.

The early investigations of the Colorado River Basin had indicated that the physical conditions of the river were not favorable for the development of cheap water power. Throughout the length of the river there were many rapids but no sheer drops, and obviously the construction of high dams would be necessary before a large amount of power could be developed. It was estimated in 1916, however, that with the construction of high dams, more than 2,000,000 horsepower might be developed without interfering with the use of water for irrigation,[70] and since then that estimate has been tripled.[71] With the improvements in construction techniques and the development of long-distance transmission for electric energy, the power possibilities of the river became very great; and plans were outlined to burden power production with the costs of flood control, river regulation, and other improvements. The fact that the gradient of the lower reaches of the river was comparatively low made possible the creation of immense reservoirs through the construction of dams in the narrow canyons, and Black Canyon gained early favor as the site best suited to begin the work of control and fuller utilization of the river by the construction of a power and storage dam. As previously discussed, the Boulder power plant, when fully equipped, will have a capacity of 1,835,000 horsepower, and will generate 4,330,000,000 kilowatt-hours of firm energy annually. The production of such a tremendous

[70] E. C. La Rue, *Colorado River and Its Utilization* (1916), pp. 169–70.
[71] *New Reclamation Era*, Vol. XVII, No. 4 (April 1936), p. 38.

quantity of electric energy is bound to give an impetus to the economic development of the Southwest and the southern coastal plain; but whether or not that impetus is warranted under the circumstances is a question upon which there was a wide difference of opinion.

One of the important points of difference was whether a real need for increased power production existed in southwestern United States. It was argued by many that there was no need for the great increase in power supply contemplated by the Boulder Canyon Project. The plan called for the sale of most of the power in southern California, and it was stated that there was no market for so much power in that area. The market would have to be nearly doubled in order to absorb all of the additional energy; and since power was already very low in price in southern California, the competition of the additional supply would be disastrous to the companies now giving adequate service in that area. A real shortage of power could not exist unless power once supplied was withdrawn or unless a new demand was created. The supply of power normally precedes the market, and there must be a void to be filled before additional power is required. No power had been withdrawn in southern California, and the only prospect for a sudden and substantial increase in the demand was the pumping requirement of the Colorado River Aqueduct.

On the supposition that a market for the energy existed in southern California it was still considered doubtful by some that Boulder power could be sold at a price there to cover production costs. It was stated that the dam would not be a dependable source of power, since the regulation of the reservoir for flood control and water storage would create wide variations in the quantities of power produced. Because of these variations in the rate of production, the power would have to be sold at a lower price than that prevailing in the market. Also the transmission costs would be high, since it was a long way from Black Canyon to the power markets in southern California. The cost of electric energy generated by steam had fallen rapidly, and it was concluded that if additional electric power were needed the buyers would not pay more for Boulder power than for the steam-generated energy, which would probably be produced at a lower cost. In any event, the science of the production of electricity was still in the experimental stage and no one could state with certainty what production costs would be a few years in the future. As a result, it was argued, it would be impossible to find any power company, public or private, that would hazard a long-term contract for Boulder energy.[72]

It is known now as a matter of fact that power companies, both public and private, have entered long-term contracts with the federal government which will yield revenues sufficient to meet the requirements of the financial plan. Thus later events have disproved the above argument on that point; but it should be noted that the argument was never really a valid objection to the project legislation. Such an objection had been anticipated and had been forestalled by the provisions in the act requiring the execution of contracts yielding revenues sufficient to repay the cost of the project and forbidding the appropriation of money to begin construction until the contracts had been completed. Repayment of the cost from project revenues was thus practically assured before building was begun.

The argument concerning the variations in power production at Black Canyon also cannot be taken as a valid objection to the project. The extremely arid and semitropic character of the lands in the lower basin make it necessary to irrigate throughout the year, thus calling for an even flow of the river, which conforms more closely with the requirements for steady power production than a stream regulated for the seasonal needs in a northern irrigation project.[73] So far as practical in this case, an even rate of flow is desirable for flood control, irrigation, and power production; and it is obvious that this type of regulation will also benefit navigation and simplify the problem of diversion of water into the Colorado River Aqueduct and the All-American Canal. The assumption is unwarranted, therefore, that there will be wide variations in power production because of the priorities of other uses for the project.

[72] Colorado River Fact-Finding Committee of Utah, *The Power Situation* (1927), p. 6.

[73] Senate Document No. 142, 67th Congress, 2d Session, *op. cit.* (1922), pp. 4, 14.

The point of difference concerning the need for additional power and the existence of markets in southern California is a more difficult question to analyze. On behalf of the project it was said that the markets existed and that they were large enough to handle even more additional power than would be produced at Black Canyon. In addition, although southern California was the chief market it was not the only market for the power. Some 100,000 horsepower could be used in Arizona to replace steam plants, increase pump irrigation, and promote the mining of low-grade ores. Power could be provided for factories; railways could be electrified; and any excess could be used for making nitrate fertilizers.[74] Nevada was expected to take some of the power, and Utah and Colorado were also possible customers. However, most of the power would probably be sold in southern California. The best hydroelectric sites in this area had already been developed; and if the construction of power dams on the Colorado had not been forbidden temporarily by Congress, southern California interests undoubtedly would have established power rights on the river before actual construction of the Boulder Canyon Project was begun. The City of Los Angeles was especially anxious to safeguard the future growth of its population and the expansion of its industries, and representatives of the City of Pasadena announced as early as 1924 that their city was willing to pay well for a share of the power and water to be made available by the Boulder Canyon Project.[75]

In 1926 Dr. Elwood Mead, Commissioner of Reclamation, stated that in spite of the tremendous quantity of electric energy to be produced at the dam, there would probably not be enough to electrify railroads and to serve other uses not receiving a preferred status in the power distribution.[76] The power was not expected to come on the markets in one large block, but

[74] Smith, *op. cit.*, pp. 527–38.

[75] House Hearing before the Committee on Irrigation and Reclamation, 68th Congress, 1st Session, *Protection and Development of Lower Colorado River Basin*, Part 2 (1924), p. 331, Statement of Mr. William J. Carr, pp. 403–8.

[76] House Hearing before the Committee on Irrigation and Reclamation, 69th Congress, 1st Session, *Colorado River Basin*, Part 1 (1926), Statement of Dr. Elwood Mead, p. 20.

rather to appear gradually as the various producing units were installed in the powerhouse. Thus an absorption period would be provided which would permit the development of agencies capable of appropriating the power for its more productive uses. Before 1928, records showed that the increase in use of firm power in the Southwestern power markets was at the rate of about 75,000 horsepower per year and that the installed horsepower was always at least twice the amount of the firm. On this basis the 1,835,000 horsepower capacity of the Boulder power plant would be fully used in about twelve years;[77] but since a large new market for power would be created with the completion of the Colorado River Aqueduct, and since many engineers predicted an increase in the rate of consumption, it was likely that the power would be absorbed in a much shorter time.[78] One survey estimated that even if the Boulder power were made immediately available in 1928, it could be absorbed within three or four years; but the results of this survey cannot be taken without qualification, since it assumed a smaller installed capacity than that called for in the final plans.[79] However, the Colorado River Board, headed by Major General William L. Sibert, reported that in its opinion the growing demand for electric energy in southern California would be sufficient to absorb the profitable power output of the project, and it was announced that the board had come to this conclusion only after a careful study of forecasts based upon apparently conservative data.[80]

It should be noted that the report of the Colorado River Board was based partially, at least, upon a survey made by the City of Los Angeles in 1925 and may reflect a bias of the city

[77] $1{,}835{,}000 \div 150{,}000 = 12.2$.

[78] House Hearing before the Committee on Irrigation and Reclamation, 70th Congress, 1st Session, *Protection and Development of the Lower Colorado River Basin* (1928), Appendix, Report of George W. Malone, p. 554. See also Senate Document No. 186, 70th Congress, 2d Session, *Colorado River Development* (1929), pp. 208–12.

[79] Arthur P. Davis, "Problems of the Colorado River," *The Annals of the American Academy of Political and Social Science*, Vol. CXXXV, No. 224 (January 1928), p. 123.

[80] House Document No. 446, 70th Congress, 2d Session, *Report of the Colorado River Board on the Boulder Dam Project* (1928), p. 15.

in favor of the Boulder Canyon Project, although the board undoubtedly took this possibility into consideration. The survey covered the territory immediately adjacent to the Boulder Canyon Project, including southern California, southern Nevada, and Arizona. Property values over a period of years were determined for the whole area and its several parts, and estimates were made of the annual gross production of industry through agriculture, mining, and manufacturing, the amount of electric power and water in use, and the extent of commerical activity. Through the use of trend curves, which were carefully checked, the probable normal future development in each of the several activities was determined. These quantitative results were then adjusted upward to allow for the stimulation of activities resulting from the utilization of Boulder Canyon power and water. The known results of the Los Angeles Aqueduct project were used as a guide in forecasting the influence of the Boulder Canyon Project on the Southwest, and were undoubtedly helpful as evidence to justify an optimistic upward trend. The survey provided the best information available for the Colorado Board to use in rendering its report, and, since the data was carefully analyzed, there was no reason to question the board's conclusions, in spite of the possible bias in the figures.[81]

As a matter of fact subsequent events placed the power-market situation in a somewhat different light. When the bids for electric energy were opened on October 1, 1929, it was found that two of the applicants, the City of Los Angeles and the Southern California Edison Company, had asked for all of the power to be developed at the dam and that all together applications had been received for more than three times the amount of power that would be available. At that time the optimistic forecasts made by the various surveys seemed to be justified; but later, new and important factors entered the picture. During the business depression which followed 1929, a decided flattening in the power-load curve appeared on the

[81] E. F. Scattergood, "Community Development in the Southwest as Influenced by the Boulder Canyon Project," *The Annals of the American Academy of Political and Social Science*, Vol. CXLVIII, No. 237 (March 1930), Part 2, pp. 3–4.

records of all the generating agencies. Both publicly and privately owned plants had spare capacity on hand, and the situation was in sharp contrast with the steadily increasing demand that was forecast in 1929. In addition, Hoover Dam was completed more than two years ahead of schedule, and some of the contractors were thus obliged to take power sooner than they had anticipated. Neither Arizona nor Nevada was in a position to make use of her allocation for some time; and an added burden would fall, therefore, upon the City of Los Angeles, the Southern California Edison Company, the Los Angeles Gas and Electric Corporation, and the Southern Sierras Power Company. The allotment to the Metropolitan Water District represented the total quantity of power that would be required ultimately when the aqueduct was operating at capacity; but since the entire amount of 1,500 second-feet would not be diverted at first, full use of the power privileges for pumping would also not be made. As a consequence of these events, conditions in 1934 indicated that the power requirements in southern California were being met by an installed capacity with an average load of about 30 per cent, which was not much over half the economic figure.[82] At that time the accuracy of the power surveys which had indicated the existence of enormous potential markets was in doubt. But again subsequent events changed the entire outlook. With business recovery came an increase in the markets for power, and the potential demand for additional installed capacity which had disappeared so suddenly was practically re-established by the time Boulder power was ready for delivery. The rate of growth of the demand for the power actually exceeded previous expectations, and the increasing population and industrial development in southern California, plus the potential need for power in Nevada and Arizona for mining and agricultural developments, made it clear that before long the Boulder power plant would be operating at capacity.[83]

It has been generally recognized that the measure of opportunity of human habitation in the Southwest, as well as in the

[82] *Engineering News-Record*, Vol. CXIII, No. 22 (November 29, 1934), pp. 689, 691.

[83] *The Reclamation Era*, Vol. XXVIII, No. 10 (October 1938), p. 213.

Los Angeles area, is the extent to which additional supplies of water of a permanent and reliable sort may be made available at reasonably low costs; but it is not so thoroughly appreciated that this condition prevails also with respect to electric power. Coal deposits in this section of the country are lacking or inaccessible; and oil fields, while producing abundantly now, are subject to such rapid exhaustion that they cannot constitute a permanent source of industrial power on a large scale. The progress of hydroelectric development in California has been remarkable; but the limit of this development has been nearly reached. Power production on the Colorado, however, is still in its infancy. Distance from centers of population and the presence of enormous coal deposits in the upper basin are the chief factors which have contributed to this lack of growth; but with the exhaustion of other sources of supply, the perfection of long-distance transmission, and the increase in domestic and industrial needs an impetus to the exploitation of the river's power resources may be anticipated.[84] The importance of power to the development of the Southwest becomes more obvious when it is recognized that even the water supplies for both domestic and agricultural uses are largely dependent upon electric energy. There are some districts in which the water required in excess of the natural rainfall may be secured through gravity systems; but in general it may be said that the water supply is considerably dependent upon pumping projects. As the development of this section proceeds, and the available water sources are more completely utilized, even greater dependence will have to be placed upon electric power to secure and deliver the necessary water supply.[85] Hydroelectric power will also play a leading role in industrial and commercial developments in the Southwest, since other sources of power are extremely inadequate, and, if those developments are as great as even the most conservative estimates, this section of the country will constitute a market for much more power than can ever be produced at

[84] Frank E. Weymouth, "Major Engineering Problems: Colorado River Development," *The Annals of the American Academy of Political and Social Science*, Vol. CXLVIII, No. 237 (March 1930), Part 2, p. 22.

[85] E. F. Scattergood, "Community Development in the Southwest as Influenced by the Boulder Canyon Project," *ibid.*, pp. 1–2.

Black Canyon. Business depressions may cause temporary decreases in the demand, but undoubtedly the potential markets exist. It may be concluded, therefore, that Boulder power will fill a real economic need and that its use in promoting domestic, agricultural, industrial, and commercial developments will lead inevitably to a still greater exploitation of the power resources of the Colorado River.

Even though it had been proved that markets could be found for Boulder power, the objection still remained that steam-generated power might be a more economic source of supply, in spite of the fact that the Boulder power plant was a very efficient producer of hydroelectric power. It was pointed out that during recent years the costs of production of steam power have been materially reduced; but this evidence alone is not positive proof of the superior efficiency of steam generation. It has not been definitely determined that the reduction will continue or that costs will stay at their present low level. The costs of steam-generated power have fallen chiefly because marked improvements have been made in fuel economy and because the costs of fuel oil and natural gas in southern California have been extremely low. New inventions may result in further improvements in fuel economy; but it is highly probable that with the depletion of the more accessible fields the prices of oil and natural gas will rise. In the past some producers have secured long-term contracts for oil or natural gas at very favorable prices; but many of these contracts are about to terminate and can be renewed only at a substantial increase in price. There is no doubt that the liquid oil reserves of the country are being exhausted. Soon after 1920 it was estimated that such reserves would be practically exhausted within twenty or twenty-five years; but since that time new fields have been discovered and the methods of extracting oil from the ground have been improved. However, the demand is continually increasing, and oil is definitely a limited resource. Enormous reserves are available in the oil shales, but at present no method is known which will produce oil from the shales as cheaply as it can be extracted from the ground in liquid form. It seems logical to conclude, therefore, that unless large, new fields are discovered, the price of oil will rise in the near future. Since the price of oil is the most important single factor in the cost of steam-generated power in southern California, it follows that the cost of that power probably will also increase.[86]

It was suggested that a decrease in cost of steam-generated power and a saving of oil might be effected through the use of coal for fuel or through the use of mercury vapor-steam turbines. The reserves of coal are far greater than those of oil, and coal is always available as a substitute for oil whenever economic conditions justify the change. However, transportation costs to southern California will prevent the use of coal as fuel for power generation in that area during the next fifty years or so unless the price of oil should rise higher than anticipated during that period. The ideal conditions under which to use coal for power generation include: (1) a balanced condition of the market for the various by-products, assuring their regular and complete sale at established market rates; (2) a very large scale of operation, using 10,000 to 20,000 tons of coal a day; and (3) a location near the mouth of the mine to avoid hauling charges on the coal. The third condition is one which cannot be fulfilled for southern California.[87] As far as the mercury vapor-steam turbines were concerned, engineers were generally agreed that the savings to be effected were not worth while, especially since the price of mercury was likely to rise with the resulting increase in the demand.[88]

It had been estimated that no other hydroelectric development could bring power to the southern California market at as low a unit cost as the one at Boulder Canyon, and since oil fuel was the cheapest fuel for steam-generated power in that market there would be no conservation of oil through changes in production methods in the southern California power industry unless Boulder power could be substituted for steam-generated power. The conservation of oil resources is very important from

[86] Department of the Interior, *Development of the Lower Colorado River* (1928), Report of Professor W. F. Durand, pp. 402–3.

[87] *Ibid.*, pp. 405–7.

[88] Senate Document No. 186, 70th Congress, 2d Session, *Colorado River Development* (1928), p. 56.

the economic point of view. It makes little difference whether the oil reserves will last for twenty-five, for fifty, or even for a hundred years, since undoubtedly in the long run the greatest benefits to society will be secured if oil is used for lubrication and other purposes for which it has no satisfactory substitute rather than for the production of steam-generated power which can easily be supplanted by hydroelectric energy. The development of one horsepower per year requires the consumption of about twenty-five barrels of oil as fuel. At that rate the substitution of 1,000,000 horsepower of hydroelectric energy for steam-generated power would result in a saving of some 25,000,000 barrels of oil annually, which would represent a definite national economy in general and a very notable economy with reference to the uses for oil.[89] It should be noted, however, that although the quantity of oil saved by such a substitution would be very great it would not be as large as the unqualified figures would indicate, since not all of the Boulder power would be substituted for steam-generated power. In 1934 it was estimated that 71 per cent of the energy generated in Los Angeles was hydroelectric power; hence some of the new power would merely supplant energy generated by the less efficient hydroelectric plants.[90] In addition it seems reasonable to conclude that the more efficient steam plants would remain in competition for some time with the Boulder power plant and other hydroelectric projects, at least until the rising cost of oil had reached the point where even the variable costs (excluding the fixed) were no longer being covered by the selling price of power.

On the basis of this reasoning it might be inferred that Boulder power could not be sold in the southern California market at a price which would compete with steam-generated power and at the same time repay the costs of the project, and as a consequence the project's financial plan would meet with failure. But this was a result which did not agree with the

[89] Senate Hearings before the Committee on Irrigation and Reclamation, 68th Congress, 2d Session, *Colorado River Basin*, Part 1 (1924), Statement of Dr. W. F. Durand, p. 19.

[90] *Engineering News-Record*, Vol. CXIII, No. 22 (November 29, 1934), p. 691.

careful forecasts which had been made of the power revenues.[91] In 1928 the Colorado River Board had concluded that the power which could be generated at Black Canyon would be a resource sufficient to amortize the cost of the project if the income from storage could be reasonably increased and if the capital investment could be reduced by the cost of the All-American Canal plus that part of the investment chargeable to flood control.[92] Later, under the completed power contracts, the annual revenue had been estimated at about $7,200,000 and the additional revenue from water storage at about $250,000. Even if the expected total annual revenue of some $7,450,000 erred materially on the side of optimism, there was still reason to believe that sufficient funds would be available to make the project self-liquidating.[93] It should be noted, however, that these revenues were subject to uncertainty due to possible future revisions of the contracts; but since over $17,000,000 worth of generating equipment was to be amortized in ten years,[94] and since the charge of $25,000,000 for flood control was not made obligatory upon revenues, the net receipts might fall considerably without endangering the operation of the financial plan. In addition, the allocation of nearly all of the costs to power production was open to question in view of the fact that the project was a multiple-purpose project. If the project had to be self-liquidating, and if electricity was the only vendible commodity, it followed that the power sales had to be sufficient to cover the costs of the entire project. Yet in the case of the Boulder Canyon Project, other services of great importance, such as flood control, silt control, and water storage, were to be rendered, and from the social and economic points of view, at least, consideration should have been given to these other services as a large part of the returns realized from the investment of government funds.[95]

[91] See Table VII, p. 145, above.

[92] House Document No. 446, 70th Congress, 2d Session, *Report of the Colorado River Board on the Boulder Dam Project* (1928), p. 15.

[93] Paul T. Homan, "Economic Aspects of the Boulder Dam Project," *The Quarterly Journal of Economics*, Vol. XLV, No. 2 (February 1931), pp. 196–97, 203.

[94] Wilbur and Ely, *op. cit.*, pp. 129, 141.

[95] Horace M. Gray, "The Allocation of Joint Costs in Multiple-Purpose Hydro-

In spite of the evidence in support of the Boulder Canyon financial plan, there was still some question whether or not the southern California market would secure as great an advantage in low power costs from a hydroelectric-power plant in Black Canyon as from steam plants located much closer to the terminal substations. The City of Los Angeles was to be one of the important customers for Boulder power, and thus the attitude of the city on this point was an important consideration. For many years the city had been developing a part of its own supply of electric energy in competition with the Southern California Edison Company; but, since its generative capacity was inadequate to meet all its needs, a major fraction of its current had to be purchased wholesale from the Southern California Edison Company, which almost monopolized the private-power enterprise in the southern part of the state. The average rate paid by the city for this power was 6.5 mills per kilowatt-hour, a price much in excess of production costs in the city's plants and in the company's more efficient plants and of the cost of Boulder power delivered to Los Angeles. Even if it were granted that in the most efficient steam plants power could be generated at a lower cost than hydroelectric energy could be provided, the market price would still stay high enough to cover the cost of the least efficient plant remaining in production. If the city continued to buy power from the company, therefore, it could not secure a price based upon the cost of production of the most efficient steam plants until the less efficient plants had been scrapped; and such a condition could not be expected until many years had passed. If the Boulder power plant were built, however, the city could buy power at a price at least 2.5 mills per kilowatt-hour below the average charge being imposed by the company and would no longer be dependent upon the company. In addition, the estimated price of Boulder power included a provision for amortization of the more expensive features of the Boulder Canyon Project, and the price might be lowered once amortization had been completed. Such a development would permit the power to be delivered at a cost lower than any con-

electric Projects," *The American Economic Review*, Vol. XXV, No. 2 (June 1935), pp. 227–28.

ceivable cost of power from private steam plants; while, on the other hand, if the anticipated increases in oil prices materialized, the cost of steam-generated energy would undoubtedly rise. In view of this situation, it was not surprising that the City of Los Angeles took little interest in the argument that it was uneconomical to bring Boulder power to the Pacific coast.[96]

The attitude of the city, however, was not the only consideration. If it be granted that power could be generated in an efficient steam plant at a lower cost than at Black Canyon, and delivered to the southern California market, the question arises why the city should not build its own steam plants to take advantage of the low cost of generation. As to this, aside from the enormous outlay that such a procedure would involve, one of the most important objections to the substitution of steam-generated energy for Boulder power was that the production of steam-generated energy would not increase greatly the market for its use, while the production of Boulder power would. A lowering in price would undoubtedly result in an increase in the amount taken; but in order to create the substantially larger market required a rapid growth in the population of southern California was necessary. As previously discussed, any great increase in population was dependent upon an enlarged water supply, and this condition suggested recourse to the Colorado River and the construction of generating facilities for pumping. Southern California might increase its power supply through the use of oil or natural gas for fuel; but the underlying problem was to secure the water needed to support the growing population, which in turn would create the market necessary to absorb the additional power.

The production of power in Black Canyon has been made possible by the construction of a dam already required for other purposes, and to have ignored the power possibilities of the project would have been an inexcusable waste of a valuable resource. The fact that power must bear most of the cost of the Boulder Canyon development is the result of the joint-cost situation involved in a multiple-purpose project; but in view of the

[96] Homan, *op. cit.*, pp. 184–85, 200–202.

flood-control, river-regulation, water-storage, and reclamation services to be gained, it would not have been an uneconomic procedure to conduct the project at a loss, if necessary, so long as the price charged for electric energy would cover the costs directly attributable to power production and distribution and would make some contribution to the joint costs. To have forced power to bear nearly the entire burden would have prevented power consumers from gaining all of the benefits they deserved; but such a result seemed inevitable with power the only major vendible commodity available. Southern California very definitely realized the value of the other services, however; and as far as this section of the country was concerned, the power element was only one of several features of a project which would increase its population, industries, wealth, and land values.

Southern California is not the only district expected to benefit from the power-production and other services of the Boulder Canyon Project; but because of its more advanced industrial development it is in a better position than other sections of the Southwest and the Pacific coast to take advantage of these services. The production of cheap power in Black Canyon will undoubtedly stimulate industry in both Arizona and Nevada and will make possible the more extensive use of electric power in irrigation and in mining;[97] but these benefits are not likely to appear as soon as those anticipated in southern California. This territory has been preparing itself for a period of rapid growth and industrial development, and the services to be derived from the Boulder Canyon Project are an important element in the general plan. Including the cost of the Colorado River Aqueduct, the cost of local distributing systems, the cost of transmission lines, and the payments for Boulder power to amortize a large part of the cost of that project, southern California will spend over $550,000,000 to insure this development and to place itself in a position to take immediate action on the

[97] E. F. Scattergood, "Engineering and Economic Features of the Boulder Dam," *The Annals of the American Academy of Political and Social Science*, Vol. CXXXV, No. 224 (January 1928), p. 122. See also G. E. P. Smith, *The Colorado River and Arizona's Interest in Its Development* (1922), p. 538; also *Report of Colorado River Commission of Nevada, Including a Study of Proposed Uses of Power and Water from Boulder Dam* (1935), pp. 10–17.

opportunities offered. The prospect of the expenditure of such an enormous amount subjected the community to very close scrutiny on the question of its ability to bear the heavy financial burden; but all of the studies made concluded that on the basis of present wealth and conservative forecasts of future growth there was little doubt that southern California would be able to maintain its financial solvency. As far as the Boulder power contracts were concerned, therefore, the federal government could reasonably expect the regular collection of the revenues necessary to fulfill the requirements of the Boulder Canyon Project Act.[98]

It was obvious that southern California planned to use the power and water services rendered by the Boulder Canyon Project to increase its own wealth. It was estimated that the $550,000,000 or more to be spent by the community would result in an increase in property values and revenues amounting to over $4,000,000,000. Cheap power was expected to bring new industries to that area, to expand those already established there, and to increase their efficiency. As a result the Boulder Canyon Project was frequently referred to as a raid by southern California upon the federal treasury. Actually, the federal government appears to have acted as banker in an enterprise which might be called somewhat speculative, since the repayment was dependent almost entirely upon the business conditions and the future growth and development of industry in southern California. However, the government had refused for some time to permit the exploitation of the power resources of the lower Colorado by other agencies, chiefly on the grounds that the international and interstate character of the river and the problem of flood control and reclamation required direct federal supervision of any developments on the stream. Yet it could not be expected to follow a policy of permanent sterilization of such a highly productive resource, and it seemed inevitable that the potential services of the river would be exploited some day. Since the federal government insisted upon being the exploiting

[98] E. F. Scattergood, "Community Development in the Southwest as Influenced by the Boulder Canyon Project," *The Annals of the American Academy of Political and Social Science*, Vol. CXLVIII, No. 237 (March 1930), Part 2, pp. 4–5.

agent, it is not reasonable to brand the areas benefited as sections of the country which have been unduly favored by national aid. Such an unbalanced distribution of benefits is implicit in the nature of the institution of private property, and southern California should be commended rather than condemned for making the best of the great opportunities offered.

The fact that the federal government had prevented the development of the power resources of the lower Colorado River under private initiative led to a bitter controversy between those who supported the government policy and those who argued that such developments should be left to the private power interests. Comparatively few questioned the propriety of government intervention for the regulation of an interstate and international stream, for the development of a flood-control system, or for the storage of water for domestic and agricultural uses. The arguments against the expenditure of funds on the All-American Canal or upon facilities for the storage and diversion of water for the Metropolitan Water District had little to do with the fact that they were government funds but were chiefly concerned with the wisdom of reclamation or the justice of the division of the waters between the states. On the power question, however, the policy of federal intervention was definitely the outstanding issue of the argument. The most important point involved was that power generated and distributed by public agencies in the Los Angeles area would indicate definitely the relative economy and efficiency of privately operated and publicly operated projects, and apparently the private interests feared that the results would not be favorable to them. For many years it had been assumed that public enterprise was far less efficient than private, and to prove that such an assumption might not be warranted would weaken the position of private enterprise in this important public-utility field. If the river had been left open to private exploitation, a series of low dams would probably have been built which could not have produced power at as low a unit cost as the present development in Black Canyon and which would not have been as efficient in the provision of the river-regulation and water-storage services so greatly desired. Some of the advocates of federal operation suggested that electricity should be burdened with only its separable costs in order to show that government multiple-purpose projects are much more efficient than private single-purpose power plants; but this suggestion was ignored in drafting the financial plan.[99] The power contracts, as written, are somewhat of a compromise between the opposed policies; but there is still enough public operation involved to carry on the test of the relative efficiency or inefficiency of government control. It is interesting to note again that the private interests have not objected so strenuously to federal supervision of the unprofitable (or less profitable) services of the project. It appears now that those who argued for private exploitation of the power resources of the lower river were attempting to draw a line between the fields of public and private monopoly, the position of which would be influenced more by the prospects of profit than by the consideration of public welfare. The policy of the Bureau of Reclamation on this point was that the federal government should not turn the power development over to private enterprise but should derive revenues from the power sales which would repay the project's construction cost and possibly also provide a fund to be used in financing additional developments on the river.[100] In the writer's opinion this was definitely the best course to follow in order to secure the greatest social benefits; and, since it was the financial plan eventually adopted, it is apparent that Congress was determined to keep all phases of the Boulder Canyon Project, including power, under the close supervision of the federal government.

There was probably more argument and nation-wide disagreement concerning the economic aspects of the Boulder Canyon power question than of any other phase of the project, including the bitter dispute over the division of water between the upper and lower basins of the Colorado River and the allocation of the lower-basin supply under the Boulder Canyon Project Act. Careful investigations had indicated that there was a need for the power and a market to absorb it, and that the energy could be sold at a price that would easily return the gov-

[99] Gray, *op. cit.*, p. 228.

[100] *Thirty-first Annual Report of the Commissioner of Reclamation for the Fiscal Year Ended June 30, 1932*, p. 16.

ernment's disbursements to the federal treasury. Other studies had indicated that federal development of the power possibilities of the project would be more economical and would contribute more to the social welfare than similar exploitation under private enterprise; but even under private development the power revenues would probably have justified the construction of the project from a financial point of view. That the government has kept control of the power-production facilities in order to assure reaping the greatest social benefits is an unusual step away from the traditional policy of reserving such activity to private enterprise; but, as an unusual step, it may serve as a precedent which will have far-reaching effects upon the future economic development of the power industry.

THE PROJECT AS A RELIEF MEASURE DURING A PERIOD OF DEPRESSION

Flood control, water storage, and power production were the most important services to be rendered by the Boulder Canyon Project. But of the more temporary benefits the employment of thousands of men during the period of business depression following 1929 occupied a prominent place as an economic argument for beginning construction soon after the necessary legislation had been passed. In 1930 President Hoover and Secretary Wilbur both appealed to the Bureau of Reclamation to speed up the drafting of building plans in order to provide employment for workers on the project at an earlier date, and the bureau complied with these requests.[101] Construction was begun in the spring of 1931, and engineers and contractors alike rushed the start of work, although the existing conditions were most unsatisfactory. Instead of adhering to the usual practice of having small crews prepare adequate facilities before the main body of men was put to work, hundreds of men were given jobs in 1931 even though Boulder City was still a barren desert. The inevitable result was severe hardships for those employed, especially since the summer temperatures for 1931 were twelve de-

[101] *New Reclamation Era,* Vol. XXI, No. 12 (December 1930), p. 245. See also Ray Lyman Wilbur and Elwood Mead, *The Construction of Hoover Dam* (1933), p. 49.

grees above normal. Thus the labor turnover during the early construction period was high; but the existence of widespread unemployment at that time made available an excellent grade of worker in spite of the adverse conditions.

As construction progressed and living and working conditions improved, more and more men were hired until the peak of construction activity in Black Canyon was reached during the latter half of 1934.[102] In April 1931, Six Companies, the principal contractor, opened an employment office in San Francisco; and by May some 1,000 men had been taken to the dam site to commence operations.[103] Later all hiring was done through a federal-state employment agency in Las Vegas, Nevada. In August about 1,300 men were on the contractor's payroll, and in October the number had increased to 2,100. By the end of 1931 some 3,100 men were employed and applications for work had been received from 42,000 people.[104] In April 1932 the number of people employed on the project was estimated at 4,200.

The minimum wage for construction work on the project was $4.00 per eight-hour day, and in January 1932 the average daily payroll was $13,100.[105] On December 31, 1932, the peak of employment was reached by Six Companies when 3,882 men were working for the contractor. At that time approximately 4,580 persons were being employed on the entire project; but, since other contractors had just begun (or were about to begin) their share of the work, it was estimated that an average of 4,600 men would be given employment for two and one-half years.[106] Employment continued to increase, and on June 30, 1934, there were 5,218 persons listed on the payrolls of the Bureau of Reclamation and the project contractors, and monthly gross wages reached a height of nearly $744,000.[107] The real peak of employment occurred on July 20, 1934, when 5,251 men were

[102] *The Reclamation Era,* Vol. XXVII, No. 2 (February 1937), p. 26.

[103] *New Reclamation Era,* Vol. XXII, No. 4 (April 1931), p. 80.

[104] *Engineering News-Record,* Vol. CVII, No. 26 (December 24, 1931), p. 1015.

[105] *Ibid.,* Vol. CVIII, No. 4 (January 28, 1932), p. 146. See also *The Reclamation Era,* Vol. XXIII, No. 2 (February 1932), p. 32.

[106] *The Reclamation Era,* Vol. XXIV, No. 4 (April 1933), p. 47.

[107] *Annual Report of the Secretary of the Interior for the Fiscal Year Ended June 30, 1934,* p. 29.

working.[108] By June 30, 1935, the number of project employees had been reduced gradually to 3,334,[109] and it continued to decline until it reached the comparatively small number of the permanent working force needed to operate the dam and power plant after completion.

The influence of the Boulder Canyon Project as a relief measure was not confined, however, to the hiring of men directly employed on the construction work in Black Canyon. Such an enormous project necessarily called for large quantities of construction material and equipment, much of which would have to be supplied from Eastern sources. Outside of the construction job itself, the cement mills, steel companies, and manufacturers of pipe, gates, valves, and electrical and hydraulic equipment experienced renewed activity in their fields, enabling them to give work to hundreds of men seeking employment. Statistical data indicate that for each 10 men engaged in construction, 18 were employed supplying materials. If 4,000 were taken as the average number of men employed on construction from 1931 to 1935, and if 4.1 persons were taken as the size of the average family in the United States, it may be computed that approximately 46,000 persons were fed, clothed, and housed during those four years through the project expenditures.[110] A check of government expenditures on the project showed that materials and supplies had come from every state in the Union, with California leading, and with Pennsylvania, New York, Wisconsin, Utah, Illinois, Alabama, Massachusetts, Michigan, and Ohio next in order as the states in which more than $1,000,000 was expended.[111] The large variety of materials needed made it possible to spread the work over a wide area, although the factor of transportation charges tended naturally to concentrate the market near the dam site whenever the materials were economically available in that section of the country.

The construction period was one of business activity and

[108] *The Reclamation Era*, Vol. XXVI, No. 4 (April 1936), p. 84.

[109] *Annual Report of the Secretary of the Interior for the Fiscal Year Ended June 30, 1935*, p. 52.

[110] *The Reclamation Era*, Vol. XXV, No. 4 (April 1935), p. 75.

[111] *Ibid.*, Vol. XXVIII, No. 1 (January 1938), p. 11.

prosperity for Boulder City and Las Vegas, Nevada. During 1934 and the first half of 1935, workers remitted about $80,000 a month from Boulder City to all parts of the country for dependent relatives or in payment of past debts, and in June 1935 Boulder City's postal savings deposits reached $521,000.[112] Southern California was also in an excellent position to benefit from the construction activity in Black Canyon through the stimulation of existing industries producing materials and supplies needed for the project and through the establishment of new industries in anticipation of the services to be derived from the completed project.

In addition to the work on the dam and the power plant southern California also benefited from the construction of the All-American Canal and the Imperial Dam, the Colorado River Aqueduct and the Parker Dam, and the gigantic transmission lines built to carry power to the aqueduct and to the coastal plain. Each one of these developments required the employment of a large number of men and the use of large quantities of materials and supplies. Work on the All-American Canal was begun late in 1934, and by March 1, 1935, there were 1,312 men employed on the project.[113] During 1934 the Imperial Valley had suffered from an unprecedented drought, and as a relief measure for those who had lost their crops and a part of their stock some 300 farmers and 900 head of stock were employed in the construction of a seven-mile section of the canal near Calexico. The payroll of about $10,000 per week changed the financial situation for the winter season in the drought area.[114] Construction of the Colorado River Aqueduct was started late in 1932,[115] and it was estimated that by June 1933 some 4,000 men were being employed and that 2,000 men would be given work over much of the construction period.[116] From December 1932 until early in 1938, between 8,000 and 10,000 men worked on the aqueduct and about 500 were employed on the Parker

[112] *The Reclamation Era*, Vol. XXV, No. 7 (July 1935), p. 134.

[113] *Engineering News-Record*, Vol. CXVI, No. 18 (April 30, 1936), p. 629.

[114] *Ibid.*, Vol. CXIV, No. 12 (March 12, 1935), p. 422.

[115] *Ibid.*, Vol. CIX, No. 26 (December 29, 1932), p. 795.

[116] *Ibid.*, Vol. CXIII, No. 10 (September 6, 1934), p. 312.

Dam project.[117] It is apparent, therefore, that in addition to the direct employment of men to work on the construction of various phases of the Boulder Canyon Project and the indirect employment of men to produce materials and supplies for that construction, many were employed in connection with other projects which were related to and dependent upon the Boulder Canyon Project.

In spite of the fact that the construction of the Boulder Canyon Project gave relief, directly or indirectly, to a large number of people during a period of severe depression, there is a question whether the relief given affected enough people to be considered as an important element in relation to the whole unemployment problem. The project was responsible for a wave of prosperity in Boulder City and Las Vegas, Nevada, and undoubtedly it was noticeably helpful as a relief measure in its effect upon southern California; but as far as the rest of the country was concerned the influence on employment was probably too slight or spread too thin to be easily recognized. It is not known how many people were helped by the Boulder Canyon Project and related projects; but if 46,000 experienced relief as a result of the construction in Black Canyon, 100,000 seems to be a generous estimate of the total number receiving aid directly or indirectly from the project and related developments. When this figure is compared with the 12,000,000 to 13,000,000 estimated to have been unemployed during 1931 and 1932 it appears to be rather insignificant.[118] A project which might solve 0.8 per cent of the total unemployment problem should not be overlooked; but also it should not be classified as a major factor of relief during a country-wide depression. Nor should the project be considered as a measure which will secure a desirable movement of the agricultural population away from the dust-bowl areas. It is true that the project will make possible the economic cultivation of highly productive lands, but the acreage is such a small part of the whole that no noticeable

[117] *The International Engineer,* Vol. LXXIII, No. 3 (March 1938), p. 78, and No. 5 (May 1938), p. 154.

[118] *Monthly Labor Review,* Vol. XXXVII, No. 5 (November 1933), p. 1129. See also *World Almanac and Book of Facts, 1938,* p. 58.

shift in population should be anticipated. It is obvious that the most trifling improvement in general agricultural practice would accomplish much more in agricultural productivity than all of the country's various irrigation projects put together.[119] In the writer's opinion, therefore, although the services to be derived from the Boulder Canyon Project are very great, its role as a relief measure was only a small element in the attempted solution of that problem and it should not be given a place of exaggerated importance.

SERVICES OF THE PROJECT AS A PLEASURE RESORT

At the time construction was begun, the Boulder Canyon Project was the greatest engineering development of its kind ever attempted and the building activities received world-wide attention. Not only were thousands of men employed on the project but hundreds of thousands of people visited the scene of operations in Black Canyon to observe the work in progress. The dam site became a point of interest to tourists almost as soon as construction started, and this interest increased as the work progressed. In 1931 records kept by the Boulder City Police Department and the reservation ranger service indicated that on the average 1,000 automobiles per week were driven to the Black Canyon observation point, and in 1932, some 92,000 people visited the dam.[120] During 1933, a total of 132,646 tourists in 48,322 cars were checked through the reservation gates, and in 1934 these figures were increased to 265,463 visitors in 84,805 cars. The average for 1935 was more than 1,000 visitors a day, and the count reached a total of some 400,000 for the year. On August 1, 1935, the number of people to arrive at the dam since March 1, 1932, not including those from local communities, passed the 711,000 mark, which was an average of 570 persons per day for nearly three and a half years.[121] A hotel was completed in Boulder City in December 1933, and several hotels were available in Las Vegas; but on many week-

[119] Isaiah Bowman, "The Pioneer Fringe," *Foreign Affairs,* Vol. VI, No. 6 (October 1927), p. 61.

[120] *The Reclamation Era,* Vol. XXV, No. 4 (April 1935), p. 77.

[121] *Ibid.,* No. 10 (October 1935), p. 197.

ends the accommodations were not adequate for the large number of guests. Even when construction activities were drawing to a close a steady stream of tourists arrived to inspect the gigantic works in Black Canyon and to see the world's largest artificial lake. The dam itself is a beautiful structure, and it fits into the general picture with such symmetry that visitors have found it a sight as fascinating as the more animated panorama presented during the spectacular building period. On April 1, 1937, the National Park Service started three checking stations, one on the Las Vegas highway, one on the Kingman, Arizona, highway, and one on the road to Lake Mead. During the first month recorded, 53,804 persons traveling in 19,167 cars were checked through the Nevada and Arizona gates, and 12,323 persons in 4,080 cars passed the Lake Mead station to see the beach and the boat docks.[122] In July 1938, a total of 65,690 persons, traveling in 21,091 cars, entered the area and 20,996 people in 6,319 cars visited Lake Mead. During the same month, 92 planes carrying 441 passengers, six special trains with 1,400 passengers, and 2,245 persons in 208 busses arrived at Boulder City. In 1937, there were 298,847 people who made the trip to the powerhouse via the elevators in the dam, and in 1938 the tabulations exceeded the 1937 figures,[123] only to be topped again in 1939.[124] These statistics indicated that tourist travel to the Boulder Canyon area was increasing and would undoubtedly be a permanent result of the construction of the project. As a consequence, the new recreational area established in this section would probably become a popular playground as soon as facilities to accommodate visitors were available.

The possibilities of the Boulder Canyon area as a pleasure resort were too great to pass unnoticed, and the development of the immediately adjacent territory for that purpose had become a part of the general plan for the project long before construction in Black Canyon was begun. It was not difficult to visualize the growth of a resort on the shores of a great lake situated in

[122] *The Reclamation Era,* Vol. XXVII, No. 6 (June 1937), p. 127.

[123] *Ibid.,* Vol. XXVIII, No. 10 (October 1938), p. 200.

[124] *Ibid.,* Vol. XXIX, No. 11 (November 1939), p. 294.

the heart of the desert. Except during the hot summer months the climate approached the ideal, and the region was one of healthfulness and rare scenic beauty. The lake is a clear, deep body of water surrounded by desert mountains of peculiar geological formation and occupying, in places, narrow canyons with walls rising thousands of feet above the surface of the water. When the road from Las Vegas to Kingman was completed in April 1938[125] the dam became a bridge on a transcontinental highway which will probably be a popular route for tourists to take for journeys from one side of the country to the other, and the resort will undoubtedly draw guests from the heavily populated districts in southern California, from Reno, Nevada, from Salt Lake City, Utah, and from Phoenix, Arizona, all located about a day's motor trip away from the dam. The branch railroad from Las Vegas to Boulder City which connects with the Union Pacific system will provide rail transportation for those who prefer train travel. From the points of view of location and transportation facilities, therefore, it appears certain that the resort will be well patronized not only by travelers on transcontinental journeys but also by thousands of people who live in the Southwest and on the Pacific coast.

To assure the development of the resort possibilities of the project along the lines of the established government policy, Lake Mead and the surrounding territory were designated as a wild-life refuge and a national recreational area and were placed under the supervision of the National Park Service.[126] The Reclamation Bureau continued to operate the dam and to supervise Boulder City in which the operating personnel lives, but development of the recreational potentialities of Lake Mead and of El Dorado Canyon below the dam was left to the National Park Service. The Reclamation Bureau provides tourist accommodations at Boulder City and a guide service for those who wish to inspect the dam and power plant, and the National Park Service is constructing roads and trails and operating sight-

[125] *The Reclamation Era*, Vol. XXVIII, No. 5 (May 1938), p. 91.

[126] United States President, Executive Order No. 6065, *Boulder Canyon Wild Life Refuge* (March 3, 1933), pp. 1–3. See also *United States Statutes at Large*, Vol. XLIX, Part 1, p. 1794.

seeing airplanes and boats. There will be no interference with grazing along Lake Mead or with prospecting and mining within the recreational area. Bathing, boating, and fishing will be the chief sports on the lake; and, since the water passing around the dam will be cold and clear, the river below is being developed into an excellent trout stream. It was recognized that because of the size and location, the potentialities of the Boulder Canyon Project as a pleasure resort were far greater than those of any other reservoir under government control.

As early as 1935 the Boulder Canyon reservoir was stocked with game fish, and later thousands of trout fingerlings were planted in the river below the dam. Fishing for bass soon became a popular pastime, since it was discovered that good-sized bass, believed to be the descendants of fish planted by Missouri settlers in the Virgin River Basin in 1904, were already present in the lake.[127] The first few months of the lake's existence attracted hundreds of people living near by for bathing and boating, and before September 1935 more than fifty pleasure craft were being operated on the lake. It is expected that the lake will become one of the most famous yachting spots in the world, since it is one of the few places in the United States where yachting can be enjoyed during the winter months, and because some of the most beautiful canyons of the Colorado River, which were previously accessible only to the most intrepid explorers, may now be entered safely by boat. By January 1936 the number of craft had increased to over two hundred,[128] including some commercial vessels which were placed in operation for passenger travel and for carrying farm and mine products to railroad crossings where supplies were picked up for return cargo. The passenger-boat business flourished, and in May 1938 nearly half of the people who visited the dam made either long or short trips on the lake. Some of these people camped near the lake, and many of them made use of the bathing beach which had formerly been a blistering stretch of sand called Hemenway Wash.

In 1937 a concession was granted to a hotel company to

[127] *The Reclamation Era,* Vol. XXVII, No. 12 (December 1937), p. 291.

[128] *Ibid.,* Vol. XXVI, No. 1 (January 1936), p. 7.

build hotels, cabins, and chalets for tourists and to provide them with automobile, boat, and horse transportation. The company was to pay an annual fee for these privileges, and all profits over 6 per cent were to be divided with the government.[129] Although these revenues would be relatively insignificant as compared with the revenues to be derived from the sale of power, they would indicate the degree of success of the recreational features of the project. These features are incidental services of the great works in Black Canyon; but they are services which may be developed easily into an important social influence upon this section of the Southwest. That the national park system has a beneficial social influence upon the country cannot be denied, and the addition of another great recreational area, even as an incidental service of a project, will undoubtedly result in a significant contribution to the general social welfare of the territory affected.

COLORADO RIVER BASIN INVESTIGATIONS

The Boulder Canyon Project, the first great federal project on the lower Colorado River, will soon be completed; but it is only the beginning in the development of a general plan to secure the utilization of the resources of the river. The builders of the project realized that they were laying the foundation for still other developments, and the plans adopted anticipated the construction of additional projects at some future time. The Boulder Canyon Project Act contained a provision authorizing the appropriation of $250,000 from the Colorado River Dam Fund to be used in investigations for the formulation of a more definite plan to secure further developments of the river when warranted. On August 26, 1930, the representatives of the various states interested in carrying on the investigations contemplated under the act met in Denver, Colorado, to devise a schedule of procedure, and surveys were begun early in June 1931 on the upper Green River in Wyoming.[130] Surveys were

[129] *Engineering News-Record,* Vol. CXVIII, No. 24 (June 17, 1937), p. 924.

[130] *Thirtieth Annual Report of the Commissioner of Reclamation for the Fiscal Year Ended June 30, 1931,* p. 24.

also made of the Parker-Gila project, and a rough classification was drawn of the irrigable lands in the Gila Basin. In 1932 arrangements were concluded to make aerial surveys of the Needles and Parker valleys on the Colorado River, and to select key points to be used in making studies of the shifts in the river channel. A canal survey was made northerly from the Gila Valley, and a field party was stationed at Overton, Nevada, to determine areas in that state which might be irrigated by pump from the Boulder Canyon reservoir. Investigations were carried on in the southern part of Utah to locate possible reservoir sites for the conservation of the surplus waters of the Paria and Virgin rivers and of Kanab Creek, and similar investigations were planned for the San Juan River Basin in Colorado, New Mexico, and Utah. Work was continued along these general lines in 1933 and the following years; and a large fund of information was gathered concerning the river and the lands which may some day be irrigated by its waters.[131]

In addition to authorizing the appropriation of $250,000 for survey work, the Boulder Canyon Project Act provided that after repayment had been made to the government for the cost of the project the excess revenues not otherwise allocated were to be kept in a separate fund to be expended for the future development of the Colorado River Basin. Probably no further development on a large scale will be warranted for a number of years, and by that time a fund may be available to enable the construction of new projects as prescribed by Congress without a heavy drain upon the federal treasury. The potential benefits to be derived from the Colorado River Basin are very great, and the future projects may yield even more valuable services than those expected from the present developments on the lower river. The fact that the Boulder Canyon Project has laid the foundations for the realization of these services has increased its social and economic importance far beyond the value of the services to be rendered directly, and it has pointed out the possibilities of the development of a great new productive area in the Southwest. On the basis of the investigations and forecasts of future

[131] *Annual Report of the Secretary of the Interior for the Fiscal Year Ended June 30, 1933,* pp. 35–36.

trends a more complete exploitation of the river's resources is apparently an inevitable result of the economic progress of this section of the country, and the Boulder Canyon Project should thus lead eventually to even greater developments which will represent a substantial addition to the total wealth of the nation.

SUMMARY

From the economic and social points of view, the Boulder Canyon Project is a development of great value to the nation in general and to the Southwest in particular. It has solved the problem of river regulation and flood control on the lower Colorado River; it has provided for the storage of water for agricultural and domestic uses; and it has made possible the generation of enormous quantities of electric power. In all of these cases the project is rendering services which fill a real need and which are being provided in the most economic manner available under the circumstances. The revenues to be derived from the project will repay its cost and will establish a fund for future developments on the river. Although the construction of the project during a period of business depression was not a substantial factor in relieving the general unemployment situation, it was a contribution toward the solution of that problem and undoubtedly had a beneficial effect in relieving unemployment conditions in the area near the building activities. In addition, it should be noted that the nation is in a position to gain even more from its investment in this project than the tangible services to be rendered would indicate. The creation of a new, large recreational area in the heart of the Southwest will undoubtedly have a beneficial effect upon the health and happiness of those able to take advantage of the new facilities, and the vast amount of fundamental knowledge gained about dams, canals, equipment, etc., obviously will result in lower costs and better construction on future jobs. The present development may be surpassed by other projects in size and in the value of services rendered; but it is doubtful if any other development will find greater justification from the economic and social points of view than the construction of the Boulder Canyon Project.

BIBLIOGRAPHY

ADAMS, FRANK. *Irrigation Districts in California*, California Department of Public Works, Bulletin No. 21, 1929.

AGRICULTURE, DEPARTMENT OF. *See* United States, Department of Agriculture.

ALL-AMERICAN CANAL BOARD. *The All-American Canal*, 1919.

ARIZONA.

Acts, Resolutions and Memorials.

Attorney General. *Motion for Leave to File Bill of Complaint and Bill of Complaint*, Supreme Court of the United States, October Term, 1930, *State of Arizona* v. *State of California et al.*

Colorado River Commission. *The Arizona Question*, 1928.

———. *Arizona's Rights in the Colorado River*, 1929.

———. *Reports*, 1927–1935.

Constitution, 1925.

Revised Code, 1928.

State Planning Board. Reports in Cooperation with National Resources Committee, Works Progress Administration, 1936.

ATTORNEY GENERAL OF THE UNITED STATES. *See* United States, Attorney General.

BAILEY, PAUL. *Flood Menace of the Colorado River as Affecting the Imperial-Coachella Area*, The Office of the State Engineer of California, 1927.

BAYLEY, E. A. "The Financial and Topographical Problems of the Colorado River Aqueduct Project," *The Annals of the American Academy of Political and Social Science*, Vol. CXLVIII, No. 237 (March 1930), Part 2, "Colorado River Development and Related Problems," edited by C. A. Dykstra.

BOULDER DAM SERVICE BUREAU. *Boulder Dam, Book of Comparisons*, 1937.

BOWMAN, ISAIAH. "The Pioneer Fringe," *Foreign Affairs*, Vol. VI, No. 1 (October 1927).

BRANDENBURG, F. H. "The Colorado River," *Monthly Weather Review*, Vol. XLVII, No. 5 (May 1919).

———. "Flood in the Colorado," *Monthly Weather Review*, Vol. XL, No. 6 (June 1912).

BROWN, ROME G. *The Conservation of Water Powers*, Senate Document No. 14, 63d Congress, 1st Session, 1913.

BUREAU OF RECLAMATION. *See* United States, Department of the Interior, Bureau of Reclamation.

CALIFORNIA.

Colorado River Commission. *Analysis of Boulder Canyon Project Act and Text of Colorado River Compact, Reclamation Law, Federal Water Power Act, Kinkaid Act*, 1930.

———. *The Boulder Canyon Project*, 1930.

———. *Colorado River and the Boulder Canyon Project*, 1931.

Constitution, 1931.
Department of Public Works. *Flow in California Streams*, Bulletin No. 5, 1923.
Special Message of Governor Friend W. Richardson to the Legislature in Extraordinary Session, October 22, 1926.
Statutes.
Supreme Court Decisions. *The Metropolitan Water District of Southern California* v. *J. E. Burney*, Decided April 3, 1911, 215 California Reports, pp. 582–87.
CENSUS BUREAU. *See* United States, Department of Commerce, Bureau of the Census.
CLARK, WALTER GORDON. *The Colorado River* (no date).
COLORADO.
Compiled Laws, 1921.
Laws.
COLORADO RIVER COMMISSION OF ARIZONA, *See* Arizona, Colorado River Commission.
COLORADO RIVER COMMISSION OF CALIFORNIA. *See* California, Colorado River Commission.
COLORADO RIVER COMMISSION OF NEVADA. *See* Nevada, Colorado River Commission.
COLORADO RIVER FACT-FINDING COMMITTEE OF UTAH. *See* Utah, Colorado River Fact-Finding Committee.
COLTER, FRED T. *Protest and Notification to Contenders Adverse to Arizona's Water and Power Rights in the Colorado River*, Arizona Reclamation Association, 1937.
CONFERENCE BETWEEN DELEGATES REPRESENTING CALIFORNIA, NEVADA, AND ARIZONA ON THE COLORADO RIVER. *Official Report of Proceedings*, August 17, 1925.
CONFERENCE OF GOVERNORS, COMMISSIONERS, AND ADVISERS OF THE STATES OF ARIZONA, CALIFORNIA, COLORADO, NEVADA, NEW MEXICO, UTAH, AND WYOMING ON THE COLORADO RIVER. *Partial Proceedings*, September 24, 1927.
Congressional Record. See United States, *Congressional Record.*
CUMMINGS, HOMER STILLÉ. *Selected Papers, 1933–1939*, edited by Carl Brent Swisher, 1939.
DAVIS, ARTHUR P. "Problems of the Colorado River," *The Annals of the American Academy of Political and Social Science*, Vol. CXXXV, No. 224 (January 1928), "Great Inland Water-Way Projects in the United States."
DELLENBAUGH, FREDERICK S. *The Romance of the Colorado River*, 1902.
Electrical Review, Vol. LXXVII, No. 17 (October 23, 1920).
Electrical West, Vol. LXXVII, No. 4 (October 1936).
Encyclopaedia Britannica. 14th edition, 1929.
Encyclopedia Americana. 1934.
Engineering News, Vols. LIV–LXXVI (July 1905 to December 1916).
Engineering News-Record, Vols. LXXVIII–CXXV (April 1917 to July 1940).
Engineering Record, Vols. LXIX–LXXV (January 1914 to March 1917).

FORBES, R. H. *Irrigating Sediments and Their Effects upon Crops,* University of Arizona, Agricultural Experiment Station, Bulletin No. 53, 1906.

———. *Irrigation in Arizona,* U.S. Department of Agriculture, Office of Experiment Stations, Bulletin No. 235, 1911.

———. *The River—Irrigating Waters of Arizona—Their Character and Effects,* University of Arizona, Agricultural Experiment Station, Bulletin No. 44, 1902.

FORTIER, SAMUEL, AND BLANEY, HARRY F. *Silt in the Colorado River and Its Relation to Irrigation,* U.S. Department of Agriculture, Technical Bulletin No. 67, 1928.

FREEMAN, LEWIS R. *The Colorado River, Yesterday, Today and Tomorrow,* 1923.

GEOLOGICAL SURVEY. *See* United States, Department of the Interior, Geological Survey.

GORDON, JAMES H. "Problems of the Lower Colorado River," *Monthly Weather Review,* Vol. LII, No. 2 (February 1924).

GRAY, HORACE M. "The Allocation of Joint Costs in Multiple-Purpose Hydro-Electric Projects," *The American Economic Review,* Vol. XXV, No. 2 (June 1935).

GRISWELL, RALPH L. "Colorado River Conferences and Their Implications," *The Annals of the American Academy of Political and Social Science,* Vol. CXLVIII, No. 237 (March 1930), Part 2, "Colorado River Development and Related Problems," edited by C. A. Dykstra.

GRUNSKY, C. E. *The Colorado River in Its Relation to the Imperial Valley, California,* Senate Document No. 103, 65th Congress, 1st Session, 1917.

HAMELE, OTTAMAR. "Federal Water Rights in the Colorado River," *The Annals of the American Academy of Political and Social Science,* Vol. CXXXV, No. 224 (January 1928), "Great Inland Water-Way Projects in the United States."

HEILPRIN, ANGELO, AND HEILPRIN, LOUIS. *Pronouncing Gazetteer or Geographical Dictionary of the World,* 1922.

HENRY, ALFRED J. "Salton Sea and the Rainfall of the Southwest," *Monthly Weather Review,* Vol. XXXIV, No. 12 (December 1906).

HOLMES, J. GARNETT. "Soil Survey of the Imperial Area, California," *Field Operations of the Bureau of Soils,* U.S. Department of Agriculture, 1903.

HOMAN, PAUL T. "Economic Aspects of the Boulder Dam Project," *The Quarterly Journal of Economics,* Vol. XLV, No. 2 (February 1931).

HOOVER, HERBERT. "The Case against Government Ownership of Utilities," *Engineering News-Record,* Vol. XCIII, No. 16 (October 16, 1924).

HOUSE OF REPRESENTATIVES. *See* United States, Congress, House of Representatives.

HOWARD, C. S. *Suspended Matter in the Colorado River in 1925–1928,* Water Supply Paper 636, 1930.

HOYT, JOHN CLAYTON, AND GROVER, NATHAN CLIFFORD. *River Discharge,* 1916.

INTERIOR, DEPARTMENT OF. *See* United States, Department of the Interior.

The International Engineer, Vol. LXXIII, No. 3 (March 1938); No. 5 (May 1938).

IVES, LIEUTENANT J. C. *Report upon the Colorado River of the West*, House Executive Document No. 90, 36th Congress, 1st Session, 1861.

JAMES, HENRY F. "The Salient Geographical Factors of the Colorado River and Basin," *The Annals of the American Academy of Political and Social Science*, Vol. CXXXV, No. 224 (January 1928), "Great Inland Water-Way Projects in the United States."

JESUNOFSKY, L. N. "New Irrigation Project on the Colorado," *Monthly Weather Review*, Vol. XXXIX, No. 9 (September 1911).

JOHNSON, HIRAM W. "The Boulder Canyon Project," *The Annals of the American Academy of Political and Social Science*, Vol. CXXXV, No. 224 (January 1928), "Great Inland Water-Way Projects in the United States."

KELLY, WILLIAM. "The Colorado River Problem," *Transactions of the American Society of Civil Engineers*, Vol. LXXXVIII, Paper No. 1558, 1925.

KIMBLE, ELLIS. "The Tennessee Valley Project," *The Journal of Land and Public Utility Economics*, Vol. IX, No. 4 (November 1933).

KINSEY, DON J. *The River of Destiny, The Story of the Colorado River*, Department of Water Power, City of Los Angeles, 1928.

LA PRADE, ARTHUR T. (Attorney General of Arizona). *Two Opinions of the Attorney General Furnished to the Colorado River Commission of Arizona*, 1933.

LA RUE, E. C. *Colorado River and Its Utilization*, Water Supply Paper No. 395, 1916.

———. *Water Power and Flood Control of Colorado River below Green River, Utah*, Water Supply Paper No. 556, 1925.

LAWSON, L. M. "The Yuma Silt Problem," *Reclamation Record*, Vol. VII, No. 8 (August 1916).

LEATHERWOOD, E. O. "My Objections to the Boulder Dam Project," *The Annals of the American Academy of Political and Social Science*, Vol. CXXXV, No. 224 (January 1928), "Great Inland Water-Way Projects in the United States."

LOCKSLEY, FRED, AND DANA, MARSHALL N. *More Power to You*, 1934.

MACDOUGAL, D. T., AND COLLABORATORS. *The Salton Sea, A Study of the Geography, the Geology, the Floristics, and the Ecology of a Desert Basin*, 1914.

MADDOCK, THOMAS. *Reasons for Arizona's Opposition to the Swing-Johnson Bill and Santa Fé Compact*, Arizona Colorado River Commission, 1927.

MEANS, THOMAS H., AND HOLMES, J. GARNETT. "Soil Survey around Imperial, California," *Field Operations of the Bureau of Soils*, U.S. Department of Agriculture, 1901.

MENDENHALL, WALTER C. *Ground Waters of the Indio Region, California*, Water Supply Paper No. 225, 1909.

Monthly Labor Review, Vol. XXXVII, No. 5 (November 1933).

MURPHY, EDWARD CHARLES, AND OTHERS. *Destructive Floods in the United States in 1905*, Water Supply and Irrigation Paper No. 162, 1906.

The National Encyclopedia, 1933.
NELSON, WESLEY R. "The Boulder Canyon Project," *Smithsonian Institution Report for 1935.*
NEVADA.
Colorado River Commission. *Memorandum, Boulder Dam Power, Including Letter to Ray Lyman Wilbur, Secretary of the Interior*, 1930.
———. *Memorandum, Including Resolutions by the Nevada Colorado River Development Commission and Letters to Secretary Wilbur in Connection with His Recent Announcement on the Division of Boulder Dam Power*, 1930.
———. *Report, 1927–1935, Including a Study of Proposed Uses of Power and Water from Boulder Dam.*
Compiled Laws, 1929.
Statutes.
NEW MEXICO.
Compiled Statutes, 1929.
Laws.
New Reclamation Era, Vols. XV–XXII (April 1924 to December 1931).
The New York Times, Vol. LXXX, No. 26,535 (September 18, 1930); Vol. LXXXII, No. 27,504 (May 14, 1933).
OLSON, REUEL LESLIE. *The Colorado River Compact*, 1926.
———. "Legal Problems in Colorado River Development," *The Annals of the American Academy of Political and Social Science*, Vol. CXXXV, No. 224 (January 1928), "Great Inland Water-Way Projects in the United States."
OREGON, SUPREME COURT DECISIONS. *Hough* v. *Porter*, Decided May 19, 1908, 51 Oregon Reports, pp. 318–456.
POWELL, J. W. *Canyons of the Colorado*, 1895.
READY, L. S. "Report on Meeting of League of the Southwest, 1921." Typewritten report from the files of Professor Charles David Marx, Stanford University, California.
RECLAMATION, BUREAU OF. *See* United States, Department of the Interior, Bureau of Reclamation.
Reclamation Era, The, Vols. XXIII–XXX (January 1932 to June 1940).
Reclamation Record, Vols. V–XI (January 1914 to December 1920).
Reclamation Record, The, Vols. XII–XV (January 1921 to February 1924).
ROTHERY, S. L. "A River Diversion on the Delta of the Colorado in Relation to Imperial Valley, California," *Transactions of the American Society of Civil Engineers*, Vol. LXXXVI, Paper No. 1528, 1923.
SAGER, GEORGE V. "Climatic Characteristics of the Boulder Dam Region," *Monthly Weather Review*, Vol. LXII, No. 6 (June 1934).
SCATTERGOOD, E. F. "Community Development in the Southwest as Influenced by the Boulder Canyon Project," *The Annals of the American Academy of Political and Social Science*, Vol. CXLVIII, No. 237 (March 1930), Part 2, "Colorado River Development and Related Problems," edited by C. A. Dykstra.
———. "Engineering and Economic Features of the Boulder Dam," *The Annals of the American Academy of Political and Social Science*, Vol. CXXXV, No. 224 (January 1928), "Great Inland Water-Way Projects in the United States."

Scattergood, E. F. "The Status of Boulder Canyon Power Allocations," *The Annals of the American Academy of Political and Social Science,* Vol. CXLVIII, No. 237 (March 1930), Part 2, "Colorado River Development and Related Problems," edited by C. A. Dykstra.

Seavey, Clyde L. "What the Boulder Dam Project Means to California and to the Nation," *The Annals of the American Academy of Political and Social Science,* Vol. CXXXV, No. 224 (January 1928), "Great Inland Water-Way Projects in the United States."

Senate. *See* United States, Congress, Senate.

Smith, G. E. P. *The Colorado River and Arizona's Interest in Its Development,* The University of Arizona College of Agriculture, Agricultural Experiment Station Bulletin No. 95, February 25, 1922.

Stevens, John F. *The Matter of the Colorado River,* Colorado River Commission of Arizona, 1930.

Sykes, Godfrey. "The Delta and Estuary of the Colorado River," *The Geographical Review,* Vol. XVI, No. 2 (April 1926).

Tait, C. E. *Irrigation in Imperial Valley, California, Its Problems and Possibilities,* Senate Document No. 246, 60th Congress, 1st Session, 1908.

Thomas, Franklin. "Metropolitan Water Distribution in the Los Angeles Area," *The Annals of the American Academy of Political and Social Science,* Vol. CXLVIII, No. 237 (March 1930), Part 2, "Colorado River Development and Related Problems," edited by C. A. Dykstra.

United States.

Attorney General. *Annual Report for the Year 1914.*

———. *Treaty of Guadalupe Hildago,* Opinion Rendered December 12, 1895, 21 Opinions of the Attorney General, pp. 274–83.

———. *Boulder Canyon Project Act,* Opinion Rendered December 26, 1929, 36 Opinions of the Attorney General, pp. 121–45.

———. *Contracts under Boulder Canyon Project Act,* Opinion Rendered June 9, 1930, 36 Opinions of the Attorney General, pp. 270–82.

Circuit Court of Appeals and Circuit and District Court. *Howell* v. *Johnston et al.,* Decided August 20, 1898, 89 Federal Reporter, pp. 556–60.

———. *Cruse* v. *McCauley,* Decided August 30, 1899, 96 Federal Reporter, pp. 369–75.

———. *Greeson et al.* v. *Imperial Irrigation District et al.,* Decided June 6, 1932, 59 (2d) Federal Reporter, pp. 529–33.

Code of Laws of a General and Permanent Character in Force January 3, 1935.

Congress, House of Representatives. Documents.

———. No. 2, 54th Congress, 1st Session, *Report of the Secretary of War,* Vol. II, 1895.

———. No. 79, 57th Congress, 2d Session, *First Annual Report of the Reclamation Service from June 17 to December 1, 1902.*

———. No. 44, 58th Congress, 2d Session, *Second Annual Report of the Reclamation Service,* 1902–1903.

———. No. 204, 58th Congress, 3d Session, *Use of the Waters of the Lower Colorado River for Irrigation,* 1905.

UNITED STATES (*continued*).

Congress, House of Representatives. Documents (*continued*).

———. No. 972, 61st Congress, 2d Session, *Message from the President of the United States Requesting an Immediate Appropriation of a Suitable Sum to Relieve the Situation on the Lower Colorado River*, 1910.

———. No. 504, 62d Congress, 2d Session, *Message of the President*, 1912.

———. No. 396, 63d Congress, 2d Session, *Combined Statement of the Receipts and Disbursements, Balances, etc., of the United States during the Fiscal Year Ended June 30, 1913.*

———. No. 1476, 63d Congress, 3d Session, *Letter from the Acting Secretary of the Treasury Transmitting a Communication of the Secretary of the Interior*, 1915.

———. No. 586, 64th Congress, 1st Session, *Letter from the Secretary of the Interior Transmitting the Report of General William L. Marshall*, 1916.

———. No. 605, 67th Congress, 4th Session, *Colorado River Compact*, 1923.

———. No. 676, 69th Congress, 2d Session, *Lower Rio Grande and Lower Colorado Rivers*, 1927.

———. No. 446, 70th Congress, 2d Session, *Report of the Colorado River Board on the Boulder Dam Project*, 1928.

———. No. 246, 71st Congress, 2d Session, *Judgments Rendered by the Court of Claims*, January 9, 1930.

———. No. 359, 71st Congress, 2d Session, *Report of the American Section of the International Water Commission, United States and Mexico*, 1930.

———. No. 798, 71st Congress, 3d Session, *Control of Floods in the Alluvial Valley of the Lower Mississippi*, Vol. I, 1931.

Congress, House of Representatives. Executive Documents.

———. No. 90, 36th Congress, 1st Session, *Report upon the Colorado River of the West by Lieutenant J. C. Ives*, 1861.

———. No. 166, 42d Congress, 2d Session, *Freight to Salt Lake City by the Colorado River*, 1872.

———. No. 1, 44th Congress, 2d Session, *Report of the Secretary of War*, Vol. II, 1876.

———. No. 1, 46th Congress, 2d Session, *Annual Report of the Chief of Engineers to the Secretary of War for the Year 1879*, Vol. II.

———. No. 1, 49th Congress, 1st Session, *Annual Report of the Chief of Engineers to the Secretary of War for the Year 1885*, Vol. II.

———. No. 1, 49th Congress, 2d Session, *Annual Report of the Chief of Engineers to the Secretary of War for the Year 1886*, Vol. II.

———. No. 18, 51st Congress, 2d Session, *Colorado River, Arizona*, 1890.

Congress, House of Representatives. Hearings.

———. Before the Committee on Flood Control, 66th Congress, 1st Session, *Colorado River Survey, Imperial Valley Project*, 1919.

———. Before the Committee on the Judiciary, 67th Congress, 1st Session, *Consent of Congress to Certain Compacts and Agreements be-*

tween the States of Arizona, California, Colorado, Nevada, New Mexico, Utah, and Wyoming, 1921.

———. Before the Committee on Irrigation of Arid Lands, 67th Congress, 2d Session, *Protection and Development of Lower Colorado River Basin,* 1922.

———. Before the Committee on Irrigation and Reclamation, 68th Congress, 1st Session, *Protection and Development of Lower Colorado River Basin,* 1924.

———. Before the Committee on Irrigation and Reclamation, 69th Congress, 1st Session, *Colorado River Basin,* 1926.

———. Before the Committee on Irrigation and Reclamation, 70th Congress, 1st Session, *Protection and Development of the Lower Colorado River Basin,* 1928.

———. Before the Committee on Irrigation and Reclamation, 70th Congress, 1st Session, *Regulating the Colorado River,* 1928.

———. Before the Committee on Rules, 70th Congress, 1st Session, *Boulder Dam,* 1928.

———. Before the Subcommittee of the House Committee on Appropriations, 71st Congress, 2d Session, *Second Deficiency Appropriation Bill for 1930.*

———. Before the Committee on Irrigation and Reclamation, 71st Congress, 3d Session, *Protection of Palo Verde Valley, California,* 1931.

———. Before the Committee on Rivers and Harbors, 75th Congress, 1st Session, *Columbia River (Bonneville Dam) Oregon and Washington,* 1937.

———. Before the Subcommittee on Appropriations, 75th Congress, 1st Session, *Interior Department Appropriation Bill for 1938.*

Congress, House of Representatives. Miscellaneous Document No. 12, 41st Congress, 3d Session, *Communication from Captain Samuel Adams Relative to the Exploration of the Colorado River and Its Tributaries,* 1870.

Congress, House of Representatives. Reports.

———. No. 87, 37th Congress, 2d Session, *Colorado Desert,* 1862.

———. No. 380, 45th Congress, 2d Session, *Fresh Water on Fort Yuma Desert,* 1878.

———. No. 1321, 49th Congress, 1st Session, *Fresh Water on Colorado Desert,* 1886.

———. No. 2440, 52d Congress, 2d Session, *Right of Way through Yuma Indian Reservation, California,* 1893.

———. No. 2578, 58th Congress, 2d Session, *Use of the Waters of Colorado River for Irrigation,* 1904.

———. No. 717, 66th Congress, 2d Session, *Provision for an Examination and Report on Imperial Valley and Other Lands in California,* 1920.

———. No. 1354, 66th Congress, 3d Session, *Change of Name of Grand River to Colorado River,* 1921.

———. No. 1657, 69th Congress, 2d Session, *Boulder Canyon Reclamation Project,* 1926.

———. No. 2285, 69th Congress, 2d Session, *Federal Power Commission Licenses Affecting Colorado River,* 1927.

UNITED STATES (*continued*).
Congress, House of Representatives. Reports (*continued*).
———. No. 918, 70th Congress, 1st Session, *Boulder Canyon Project*, 1928.
———. No. 2864, 71st Congress, 3d Session, *Protection of the Palo Verde Valley, California*, 1931.
———. No. 587, 72d Congress, 1st Session, *Protection of the Palo Verde Valley, California*, 1932.
———. No. 1281, 72d Congress, 1st Session, *Tax Laws of Nevada and Arizona to Apply to Boulder Dam*, 1932.
Congress, Senate. Documents.
———. No. 212, 59th Congress, 2d Session, *Message from the President of the United States Relative to the Threatened Destruction by the Overflow of the Colorado River in the Sink or Depression Known as the Imperial Valley or Salton Sink Region*, 1907.
———. No. 246, 60th Congress, 1st Session, *Irrigation in Imperial Valley, California, Its Problems and Possibilities*, 1908.
———. No. 357, 61st Congress, 2d Session, *Treaties, Conventions, International Acts, Protocols and Agreements between the United States of America and Other Powers, 1776–1909*, Vol. 1.
———. No. 846, 62d Congress, 2d Session, *Flood Waters of the Colorado River*, 1912.
———. No. 847, 62d Congress, 2d Session, *Colorado River*, 1921.
———. No. 14, 63d Congress, 1st Session, *The Conservation of Water Powers*, 1913.
———. No. 232, 64th Congress, 1st Session, *Imperial Valley, California*, 1916.
———. No. 246, 64th Congress, 1st Session, *Power of the Federal Government over Development and Use of Water Power*, 1916.
———. No. 103, 65th Congress, 1st Session, *The Colorado River in Its Relation to the Imperial Valley, California*, 1917.
———. No. 142, 67th Congress, 2d Session, *Problems of Imperial Valley and Vicinity*, 1922.
———. No. 186, 70th Congress, 2d Session, *Colorado River Development*, 1929.
———. No. 125, 72d Congress, 1st Session, *Flood Protection, Palo Verde, California*, 1932.
Congress, Senate. Executive Document No. 81, 32d Congress, 1st Session, *Report of the Secretary of War*, 1852.
Congress, Senate. Hearings.
———. Before the Committee on Irrigation and Reclamation, 68th Congress, 2d Session, *Colorado River Basin*, 1925.
———. Before the Committee on Irrigation and Reclamation, 69th Congress, 1st Session, *Colorado River Basin*, 1925.
———. Before the Committee on Irrigation and Reclamation, 70th Congress, 1st Session, *Colorado River Basin*, 1928.
Congress, Senate. Reports.
———. No. 5545, 59th Congress, 2d Session, *Interests of United States on Lower Colorado River, etc.*, 1907.

———. No. 1066, 67th Congress, 4th Session, *Southern Pacific Company,* 1923.
———. No. 654, 69th Congress, 1st Session, *Boulder Canyon Reclamation Project,* 1926.
Congressional Record.
Constitution.
Department of Agriculture, Office of Experiment Stations.
———. Bulletin No. 158, *Annual Report of Irrigation and Drainage Investigations,* 1904.
Department of Agriculture, *Report of the Secretary,* 1927, 1933.
Department of Commerce, Bureau of the Census. *Fifteenth Census of the United States, 1930, Population,* Vol. I.
Department of the Interior. *Annual Reports,* 1927, 1933–1937.
Department of the Interior, Bureau of Reclamation.
———. *Annual Reports,* 1902–1904, 1910, 1916–1922, 1926, 1929–1932.
———. *Boulder Canyon Project, General Regulations for Lease and Purchase of Power, Hoover Dam,* 1931.
———. *Boulder Canyon Project, Rate which Public and Private Corporations Can Afford to Pay for Power at Boulder Canyon and Rate which Will Produce Sufficient Revenue to Repay Cost of Boulder Canyon Dam and Power Plant in Fifty Years with Interest,* 1929.
———. *Boulder Canyon Project—Questions and Answers,* 1933, 1936.
———. *Boulder Dam* (no date).
———. *Construction of Boulder Dam* (no date).
———. *Dams and Control Works,* 1938.
———. *Development of the Colorado River,* 1928. (Reports by Special Advisers to the Secretary of the Interior—Hon. F. C. Emerson, Governor of Wyoming; Professor W. F. Durand, Stanford University; Hon. J. G. Scrugham, former Governor of Nevada; Hon. James R. Garfield, former Secretary of the Interior.)
———. *General Information Concerning Boulder Canyon,* 1933.
———. *General Information Concerning the Boulder Canyon Project,* 1936.
———. *General Information, Grand Coulee Dam — Columbia Basin Project, Washington,* 1938.
———. *Grand Coulee Dam, The Columbia Basin Reclamation Project* (no date).
———. *President Franklin Delano Roosevelt Visits Boulder Dam,* 1935.
———. Specifications No. 519, *Specifications, Schedule and Drawings, Hoover Dam, Power Plant and Appurtenant Works, Boulder Canyon Project, Arizona—California—Nevada* (no date).
Department of the Interior, Geological Survey.
———. *Map of the Colorado River Basin,* 1914.
———. Water Supply and Irrigation Paper No. 166, *Report of Progress of Stream Measurements for the Calendar Year 1905.*
———. Water Supply Paper No. 345, *Contributions to the Hydrology of the United States,* 1914.
———. Water Supply Paper No. 549, *Surface Water Supply of the United States,* Part 9, Colorado River Basin, 1922.

UNITED STATES (*continued*).
Department of the Interior, Geological Survey (*continued*).
———. See also Water Supply Papers Nos. 85, 100, 133, 177, 213, 249, 269, 289, 309, 329, 359, 389, 409, 439, 459, 479, 509, 529, 549, 569, 589, 609, 629, 649, 669, 689, 704, 719, 734, 749, 764, 789, 809.
Federal Power Commission. *Annual Reports*, 1921–22, 1929–31, 1933.
———. *Uses of the Upper Columbia River*, 1925.
———. *Water Powers of California*, 1928.
Geological Survey. *See* United States, Department of the Interior, Geological Survey.
House of Representatives. *See* United States, Congress, House of Representatives.
President. Executive Order No. 6065, *Boulder Canyon Wild Life Refuge*, March 3, 1933.
Revised Statutes Passed at the 43d Congress, 1st Session, 1873–74.
Senate. *See* United States, Congress, Senate.
Statutes at Large.
Supreme Court. *State of Arizona, Complainant*, v. *State of California, State of Nevada, State of Utah, State of New Mexico, State of Colorado, State of Wyoming, and Ray Lyman Wilbur, Secretary of the Interior, Defendants.* Motion for Leave to File Bill of Complaint and Bill of Complaint. October Term, 1930.
———. *State of Arizona, Complainant*, v. *State of California, State of Nevada, State of Utah, State of New Mexico, State of Colorado, State of Wyoming, and Ray Lyman Wilbur, Secretary of the Interior, Defendants.* Brief of Complainant in Opposition to Motions to Dismiss the Bill of Complaint. October Term, 1930. No. 19, Original.
———. *State of Arizona, Complainant*, v. *State of California, et al., Defendants.* Return on Behalf of the State of California, Palo Verde Irrigation District, Imperial Irrigation District, Coachella Valley County Water District, the Metropolitan Water District of Southern California, City of Los Angeles, City of San Diego, and County of San Diego to the Rule to Show Cause, Dated February 20, 1934, and Brief in Support of Return to the Rule. October Term, 1933.
Supreme Court. Decisions.
———. *McCulloch* v. *State of Maryland et al.*, Decided March 7, 1819, 4 Wheaton (U.S. Reports), pp. 159–213 (Par. 316–437).
———. *Merrit Martin and Others* v. *The Lessee of William C. Waddell*, Decided January, 1842, 16 Peters (U.S. Reports), pp. 234–79 (Par. 367–434).
———. *John Polland et al.* v. *John Hagan et al.*, Decided January, 1845, 3 Howard (U.S. Reports), pp. 238–65 (Par. 212–35).
———. *The Propeller Genesee Chief et al.* v. *Fitzhugh et al.*, Decided December, 1851, 12 Howard (U.S. Reports), pp. 471–95 (Par. 443–65).
———. *Irvine* v. *Marshall et al.*, Decided December, 1857, 20 Howard (U.S. Reports), pp. 558–71.
———. *The Daniel Ball*, Decided December, 1870, 10 Wallace (U.S. Reports), pp. 557–66.

———. *Gibson* v. *Chouteau,* Decided December, 1870, 13 Wallace (U.S. Reports), pp. 92–104.

———. *The Montello,* Decided October, 1874, 20 Wallace (U.S. Reports), pp. 430–45.

———. *Barney* v. *Keokuk,* Decided October, 1876, 94 U.S. Reports, pp. 324–42.

———. *Camfield* v. *United States,* Decided May 24, 1897, 167 U.S. Reports, pp. 518–28.

———. *St. Anthony Falls Water Power Company* v. *St. Paul Water Commissioners,* Decided November 29, 1897, 168 U.S. Reports, pp. 349–74.

———. *Green Bay & Mississippi Canal Company* v. *Patten Paper Company,* Decided November 28, 1898, 172 U.S. Reports, pp. 58–82.

———. *Green Bay & Mississippi Canal Company* v. *Patten Paper Company,* Decided February 20, 1899, 173 U.S. Reports, pp. 179–90.

———. *Kansas* v. *Colorado,* Decided May 13, 1907, 206 U.S. Reports, pp. 46–118.

———. *Scott* v. *Lattig,* Decided February 3, 1913, 227 U.S. Reports, pp. 229–45.

———. *United States* v. *Midwest Oil Company,* Decided February 23, 1915, 236 U.S. Reports, pp. 459–512.

———. *Economy Light & Power Company* v. *United States,* Decided April 11, 1921, 256 U.S. Reports, pp. 113–25.

———. *Wyoming* v. *Colorado,* Decided June 5, 1922, 259 U.S. Reports, pp. 419–97.

———. *Arizona* v. *California et al.,* Decided May 18, 1931, 283 U.S. Reports, pp. 423–64.

———. *Arizona* v. *California et al.,* Decided May 21, 1934, 292 U.S. Reports, pp. 341–60.

———. *United States* v. *Arizona,* Decided February 11, 1935, 294 U.S. Reports, p. 695.

———. *United States* v. *Arizona,* Decided April 29, 1935, 295 U.S. Reports, pp. 174–92.

———. *Arizona* v. *California et al.,* Decided May 25, 1936, 298 U.S. Reports, pp. 558–72.

———. *Arizona* v. *California et al.,* Decided October 12, 1936, 299 U.S. Reports, p. 618.

War Department. *Annual Report,* 1936.

———. *Annual Reports of the Chief of Engineers to the Secretary of War,* 1879, 1885, 1886.

———. *Reports of the Secretary,* 1852, 1876, 1895.

War Department, Corps of Engineers. *Improvement of Columbia River at Bonneville, Oregon,* 1938.

UTAH.

Colorado River Fact-Finding Committee. Bulletin No. 2, *Protection to the Imperial Valley,* September 1, 1927.

———. Bulletin No. 3, *Water for Los Angeles—The Power Situation,* October 15, 1927.

UTAH (*continued*).

Colorado Fact-Finding Committee (*continued*).

———. Bulletin No. 4, *Irrigation—Reclamation—"All-American" Canal*, December 24, 1927.

Laws.

Revised Statutes, 1933.

VOSKUIL, WALTER H. *The Economics of Water Power Development*, 1928.

WARD, CHARLES B. *Explanation of Terms in the Colorado River Controversy between Arizona and California*, Arizona Colorado River Commission, 1929.

WATER SUPPLY AND IRRIGATION PAPERS. *See* United States, Department of the Interior, Geological Survey.

WEEKS, DAVID, AND WEST, CHARLES H. *The Problem of Securing Closer Relationship between Agricultural Development and Irrigation Construction*, University of California, College of Agriculture, Agricultural Experiment Station Bulletin No. 435, September, 1927.

WEYMOUTH, FRANK E. "Major Engineering Problems: Colorado River Development," *The Annals of the American Academy of Political and Social Science*, Vol. CXLVIII, No. 237 (March 1930), Part 2, "Colorado River Development and Related Problems," edited by C. A. Dykstra.

———. *Report on the Problems of the Colorado River*, U.S. Department of the Interior, Bureau of Reclamation, 1924.

WHITE, SAMUEL. *Memorandum of Law Points and Authorities Respecting the Rights of Arizona in the Colorado River*, 1925.

WHITNEY, MILTON, AND OTHERS. *Field Operations of the Bureau of Soils, 1901*, U.S. Department of Agriculture.

———. *Field Operations of the Bureau of Soils, 1903*, U.S. Department of Agriculture.

WILBUR, RAY LYMAN, AND ELY, NORTHCUTT. *The Hoover Dam Power and Water Contracts and Related Data*, 1933.

WILBUR, RAY LYMAN, AND MEAD, ELWOOD. *The Construction of Hoover Dam—Preliminary Investigations, Design of Dam, and Progress of Construction*, 1933.

WILLIAMS, ED. F., SCOTT, GEORGE W., AND HAUSER, L. A. "Palo Verde Irrigation District, California," House Hearings before the Committee on Irrigation and Reclamation, 71st Congress, 3d Session, *Protection of Palo Verde Valley, California*, 1931.

The World Almanac and Book of Facts, 1938.

WYOMING.

Laws.

Revised Statutes, 1931.

INDEX